AN INTRODUCTION TO A THEORY OF FIELDS

A simplified and extended account of a new fundamental theory of physics

I W Mackintosh

Published by New Generation Publishing in 2015

Copyright © Winstone Research Limited 2015

First Edition

The author asserts the moral right under the Copyright, Designs and Patents Act 1988 to be identified as the author of this work.

All Rights reserved. No part of this publication may be reproduced, stored in a retrieval system or transmitted, in any form or by any means without the prior consent of the author, nor be otherwise circulated in any form of binding or cover other than that which it is published and without a similar condition being imposed on the subsequent purchaser.

www.newgeneration-publishing.com

ISBN: 978-1-78507-401-1

New Generation Publishing

By the Author

A Theory of Fields

First published 2010
Authors OnLine

Second Edition 2015
Published by New Generation Publishing

ISBN: 978-1-78507-373-1

The Author

I W Mackintosh read Natural Sciences at King's College, Cambridge, specialising in Physics. After graduation he joined the Scientific Civil Service and spent his career in defence research, working on solid state lasers, solid state microwave devices and their system applications, and research on large distributed systems. He held a succession of senior posts managing then directing research. He left his final post in Whitehall and set up his own consultancy company when he also started work on 'A Theory of Fields'. He is married to Sue and they have three grown up children.

Preface

My Book 'A Theory of Fields' was published in 2010. It proposes a new fundamental theory of physics. It is based on only two postulates from which much of modern physics can be derived.

In order to increase the attention paid to this new theory I decided that a simplified account was required bringing out the main principles and conclusions. In the meantime I had extended the theory, covering many of the aspects of 'The Way Ahead' at the end of the first book (Chapter 14) and so a second book provided the opportunity to include the new material. This book provides a simplified and extended account of a new fundamental theory of physics.

My thanks are to Dr Mike White for his comments, criticisms and suggestions for the opening chapters, also to my son Peter who has urged me on with this book. However any errors in the book are mine. There is no guarantee that the mathematics are correct. You have been warned – this is speculative physics. However the ideas are new and I believe are worthy of people's attention.

Contents

Preface
Introduction
Summary of the previous Book
Symbols and units

Chapter 1   A connection between electricity and gravity   1
1.1   Introduction
1.2   The postulates
1.3   The Field Equations
1.4   The Field Equations and the development of the new theory
1.5   Summary

Chapter 2   A model for fundamental particles   8
2.1   Introduction
2.2   The particle model
2.3   Particle internal fields
2.4   Particle categories
2.5   Particle fields for a spherical particle
2.6   The effect of the source particle external steady state densities on target particles
2.7   The external potentials
2.8   The background
2.9   Summary

Chapter 3   Photons   23
3.1   Introduction
3.2   Omega waveforms
3.3   The connection between the external potentials and the omega waveforms
3.4   Photons
3.5   Photon transport of radiation energy and gravitational mass
3.6   Summary of proton properties
3.7   Summary

Chapter 4 Classical and quantum mechanics   40
4.1 Introduction
4.2 The distributed particle
4.3 The introduction of Planck's constant
4.4 The classical equations of energy and motion
4.5 Schrödinger's equation
4.6 Summary

Chapter 5 Gravitation   50
5.1 Introduction
5.2 Motion in a gravitational field
5.3 The effect of the gravitational potential on orbital motion
5.4 The gravitational red shift
5.5 Bending of light in a gravitational field
5.6 The status of general relativity within the new theory
5.7 Conclusions
5.8 Summary
Chapter 5 references

Chapter 6 Baryons and mesons   59
6.1 Introduction
6.2 Particle parameters
6.3 The spherical particle shell model
6.4 The thickness of the shells
6.5 The proton
6.6 Expressions for the constants A and K
6.7 Anti-protons
6.8 Neutral particles
6.9 Mesons
6.10 Summary

Chapter 7 Leptons   75
7.1 Introduction
7.2 Standing wave solutions
7.3 A model for the electron
7.4 The positron
7.5 The muon
7.6 Neutrinos
7.7 Summary

Chapter 8   Constants and particle parameters   83
8.1   Introduction
8.2   Fundamental constants
8.3   The electronic charge
8.4   The volume of the proton
8.5   The constants A_m and A_ρ
8.6   The constants A and K
8.7   The gravitational constant
8.8   Particle structures and parameters
8.9   Summary
Chapter 8 reference

Chapter 9   Proton and neutron structures   93
9.1   Introduction
9.2   Structure of the proton
9.3   Structure of the neutron
9.4   Summary

Chapter 10   The nuclear force   105
10.1   Introduction
10.2   The overlap model
10.3   The overlap potential energies
10.4   Summary
Chapter 10 reference

Chapter 11   Anti-particles   115
11.1   Introduction
11.2   Anti-particles with negative gravitational mass
11.3   An alternative model for anti-particles with positive gravitational mass
11.4   Conclusions
11.5   Summary
Chapter 11 reference

Chapter 12   The origin of probability in quantum mechanics 120
12.1   Introduction
12.2   Probability and hidden variables
12.3   Probability, scattering and other processes
12.4   Measurement
12.5   The postulates of quantum mechanics
12.6   Summary

Chapter 12 references

Chapter 13 Quantum field theory and quantum electrodynamics   129
13.1   Introduction
13.2   Relativistic quantum mechanics
13.3   Quantum mechanics and the basis for quantum field theory
13.4   Particle description, symmetries and the standard model
13.5   Quantum electrodynamics
13.6   QED treatment of the Lamb shift
13.7   The new theory's electron model in the treatment of the Lamb shift
13.8   Conclusions
13.9   Summary
Chapter 13 references

Chapter 14 Composite particles   139
14.1   Introduction
14.2   The x-energy diagram and the nucleon single omega solutions
14.3   The x-energy diagram and independent Dirac particles
14.4   Composite particles and an excited state of the proton
14.5   Kaons
14.6   Pions
14.7   Summary
Chapter 14 references

Chapter 15 Quarks and the structure of hadrons   156
15.1   Introduction
15.2   Quarks and the Dirac particle content of nucleons, kaons and pions
15.3   Gluons
15.4   Quark production in e^+e^- collisions at around 940 MeV and above
15.5   The volume lines on the x-energy diagrams
15.6   The phi meson
15.7   The charm quark and charmonium
15.8   The bottom quarks
15.9   Summary of particle predictions
15.10   The scattering ratio R
15.11   The connection between the new theory and the standard model

15.12 Summary
Chapter 15 references

Chapter 16   Weak processes   181
16.1 Introduction
16.2 Beta decay
16.3 W and Z particles
16.4 Summary
Chapter 16 references

Chapter 17   The background   187
17.1 Introduction
17.2 The background
17.3 Electrons and electron neutrinos
17.4 Charge densities due to distant objects
17.5 Summary
Chapter 17 references

Chapter 18   The origin of the postulates and the fundamental constants   195
18.1 Introduction
18.2 Something out of nothing
18.3 The proposed pure logic sequence
18.4 Pure logic versus conjecture
18.5 Fundamental constants
18.6 Summary
Chapter 18 references

Chapter 19   The way ahead   201
19.1 Introduction
19.2 Accuracy of the predictions of constants and parameters
19.3 Quantum mechanics
19.4 Quantum field theory and quantum electrodynamics
19.5 Nuclear theory
19.6 More particles
19.7 Particle decays and interactions
19.8 Electroweak theory
19.9 General relativity
19.10 The background and cosmology
19.11 Fundamental constants
19.12 Tests against experimental data

Chapter 19 references

Appendix A The discontinuities in charge density 206

Appendix B The omega and photon waveforms 211
B.1　Introduction
B.2　The solution of the Field Equations and the omega waveforms
B.3　Photon waveforms in the background
B.4　Photon waveforms in particles
B.5　Conclusion
B.6　Comment on Figure 3.8

Appendix C The velocity of muon neutrinos 220
C.1　Introduction
C.2　Electromagnetic radiation and particle motion
C.3　Situation with the clock in a gravitational potential Φ
C.4　Standard clock in free fall
Appendix C references

Appendix D The structure of the proton and related topics 226
D.1　Introduction
D.2　Spherical and cylindrical shells
D.3　Revised parameters
D.4　The steady state gravitational mass and charge in spherical and cylindrical shells
D.5　Equations for the particle sub-set

Appendix E The nucleon overlap potential energies 241
E.1　Introduction

Appendix F Derivation of the postulates of quantum Mechanics 250
F.1　Introduction
F.2　Postulate 1
F.3　Postulate 2
F.4　Postulate 3
F.5　Postulate 4
F.6　Postulate 5
F.7　Postulate 6
F.8　Postulate 7

Appendix F references

Appendix G Composite particles and kaons   261
G.1   Introduction
G.2   Composite particles and the subset
G.3   Kaons
G.4   Charged kaons
G.5   Neutral kaons
G.6   The inertial masses of particles and their antiparticles

Appendix H $q\bar{q}$ production in e^+e^- collisions   273
H.1 $q\bar{q}$ production in e^+e^- collisions
Appendix H references

Appendix I Models for the *W* and *Z* particles   276
I.1   Introduction
I.2   *W* and *Z* particle models analysis

Appendix J Types of particle   282
J.1   Introduction
J.2   Particles in the standard model
J.3   Particles in the new theory
J.4   Types of structure
J.5   Types of interaction
J.6   The definition of the subset

Appendix K The formalism sequence   288

List of symbols   290

Index   300

Introduction

Modern physics is extremely successful in accounting for many of the mechanisms and processes underlying the world that we see around us. However it is compartmentalised into a number of separate areas, for example classical mechanics, electromagnetism and so on, and it has been the ambition of many investigators over many years to find a unifying theory underpinning these separate areas. My book 'A Theory of Fields' was published in 2010. It offers a new approach to a unifying fundamental theory of physics. From now on I shall refer to this book as the Book.

This present book gives a simplified account of the new theory. By keeping the aim clear, we wish to expose how the laws and postulates of conventional physics arise from the new theory as quickly as possible. It will be seen by the end of the fourth chapter that the fundamentals of classical mechanics, quantum mechanics and electromagnetism are derived. Special relativity is also derived, but relativistic detail has been omitted from this account. Thus by the end of Chapter 4 a lot of the physics underpinning mechanical engineering, electrical and electronic engineering, chemical engineering, structural engineering, solid state applications, applied physics based on atomic theory and theoretical chemistry, and much more, follows from the new theory.

This present book is aimed at those who have found or will find the Book hard to follow, either because of my exposition or the complexity of the mathematics or the sheer volume of detail. The approach in the present book is illustrative not comprehensive. If the reader wants to delve into detail or see how a particular development in the theory is derived, they will need to refer to the Book. The mathematics has been kept to a minimum, both in the amount and complexity and by avoiding the use of advanced notation. So by and large this book provides simple formulae, though on a few occasions something more complicated will appear. Anyway an attempt has been made to give an account which can be understood without understanding the equations.

This book is aimed at a range of readers which includes professional physicists, undergraduates, and interested laymen. They will have a range of mathematical skills, from the professional mathematician down to school algebra. I ask that those who do not have advanced mathematics to be persistent – hopefully enough narrative has been added to make the material intelligible; and I ask the professional theoretical physicist to be patient; the purpose is to present the major principles and results, and not to obscure them in a mathematical fog.

So the purpose of this present book is to provide a simplified account of the new theory. Chapters 1 to 10 parallel Chapters 1 to 10 in the Book. Chapters 1 to 10 of this present book make many simplifications, and focus on the main results without descending into detail or ramifications. Each chapter of the Book reviews previous work which provides the background for each endeavour of the new theory. As a consequence there is not the need to reference the previous work in this book – the references are already in the Book. So the topics in these chapters in this book give an outline account of the topics that are dealt with in detail in the corresponding chapter in the Book.

In Chapters 11 to 19 opportunity is taken to extend the development of the new theory to probability in quantum mechanics, to include quantum field theory and quantum electrodynamics, and to develop a theory of quarks. Supporting detail is placed in the appendices. Whereas the main chapters of this book can be read as standalone, the appendices rely heavily on the detailed results from the Book.

An important concept is that of a formalism. A formalism is the development of a subject area from a set of laws or postulates without making further assumptions, and which leads to the major principles and additional detail of the subject area. Occam is well known for the maxim 'entities are not to be multiplied without necessity' which Russell (1961 pp 462-463) interprets to mean that if everything in some science can be interpreted without assuming this or that hypothetical entity, there is no reason to assume it. We can then argue by applying Occam's razor that the fewer the number of laws and postulates in a particular area is for the better. We can also say that the fewer the number of fundamental constants is for the better.

The major and separate compartments of conventional physics are based on their individual laws and postulates, and textbooks set out these subjects as formalisms. These compartments are:

(1) Classical mechanics, based on Newton's laws of motion

(2) Electromagnetism. It is sufficient to base it on Coulomb's law, Ampere's law, Faraday's law of induction and the concept of Maxwell's displacement current
(3) Gravity, based on Newton's law of gravitation
(4) Special relativity, based on Einstein's special relativity principle
(5) General relativity, based on Einstein's principles
(6) Quantum mechanics. Appendix F examines the postulates of quantum mechanics in detail. Quantum mechanics leads to quantum field theory and quantum electrodynamics
(7) The standard model of particle physics postulates the existence of quarks, leptons and the quanta of the various forces, i.e. photons, gluons, $W^{\pm}$ and Z particles

There are other areas which are derived from these compartments which include statistical physics, thermodynamics, nuclear physics, atomic physics, molecular physics, solid state physics, fluid mechanics, etc. From the compartments and these other areas, results are obtained which are applied to and underpin technology and engineering.

Just as the compartments are formalisms, then so is the new theory. It is based on just two postulates. The objective is now clear. The aim is to develop a formalism from the two postulates of the new theory and hence derive the laws and postulates of the compartments of conventional physics. In so doing we need to account for the required fundamental constants. Since the formal development is now contained in two books, Appendix K traces the formal thread of the development of the new theory in the two books. From the perspective of Occam's razor, a theory of fundamental physics based on just two postulates and five fundamental constants is quite a feat. Towards the end of the book, in Chapter 18, we explore the possibility that the theory might arise from no laws at all, and that there are no fundamental constants but instead just a set of constants with fixed numerical values. This is the ultimate in the application of Occam's razor.

It may be said that this book should review the whole of modern physics, so that it can be seen what the new theory achieves in prediction, explanation and resolution of problems. This would make the present book voluminous and intractable, and it would never be completed. There are many popular accounts and text books of varying difficulty which review modern physics admirably. You may come to this book with a knowledge of conventional physics and can easily separate the unconventional from the conventional. Nevertheless you have been warned – this is an account of unconventional physics.

By simplification and keeping the aim clear, we wish to expose how the laws and postulates of conventional physics arise from the new theory as quickly as possible. Chapter 1 sets out the starting point by stating the two postulates on which the new theory is based.

Reference to the Introduction

Russell B 1961 *History of western philosophy* George Allen and Unwin

Summary of the previous Book

It is not necessary for the reader to have read the previous Book ('A Theory of Fields') – the detail in it is for reference by the specialist – but a summary of its content is provided below. This is the material which is presented in a simpler and more direct form in Chapters 1 to 10 in the present book.

The new theory in the Book leads to models for fundamental particles, a model for photons and to the origins of classical and quantum mechanics. This new theory is in stark contrast to the theories which assume general relativity and quantum theory from the outset. Quantum mechanics arises from the new theory without assuming it. The predictions of general relativity (at least with regard to phenomena within the solar system) emerge from the new theory without having to introduce the assumptions of general relativity or the general relativistic concept of distortions in space-time. These conclusions can be recast to allow general relativity to be introduced as an alternative formulation within the new theory. The new theory is based on a three dimensional space together with time, with no appeal whatsoever to higher dimensional spaces. The new theory is based on two postulates and five fundamental constants. The development of the formalism leads to a set of Field Equations which provide the basis for all the subsequent work.

A general fundamental particle model is developed by constructing appropriate solutions of the Field Equations. The resulting particles are not confined to points - each has a structure occupying a volume in space. The particles' internal fields are composed of oscillatory and steady state components. The oscillatory fields extend beyond the particle boundaries into the surrounding space where they give rise to the potentials generated by the particles.

Travelling wave solutions of the Field Equations lead to the existence and properties of photons. Solutions of the Field Equations are also obtained in the presence of electric and gravitational potentials. Classical and quantum mechanics together with Newton's law of gravitation and Coulomb's law emerge from the Theory. An expression is obtained for Planck's constant.

By refining the proton and neutron models, components of the internal structures are identified corresponding to the experimentally confirmed quarks. The theory predicts the quark charge fractions. A model is proposed for the force between nucleons leading to the prediction of the interaction potential between neutrons and neutrons, protons and protons, and neutrons and protons.

Models are proposed for the proton, neutron, kaon, pion, muon, electron and electron neutrino, leading to predictions of their shapes, sizes, inertial masses and their internal structures. The values of the electronic charge and the gravitational constant are derived from first principles. It makes many detailed predictions, providing many opportunities to test the new theory experimentally. This is particularly the case with the descriptions of the internal structures of fundamental particles which are on a scale that can be probed experimentally.

In the present book there are a number of areas where the presentation and development in the Book have been improved or errors corrected. These include Chapter 3 (correction to a background factor), Chapter 5 (the derivation of the planetary orbital equation of energy), Chapter 14 (excited states of the proton), Chapter 17 (corrections to calculation of charge densities induced by distant objects) and in Appendix E (corrections to calculations of potential energies in nucleon – nucleon scattering).

Symbols and units

There is a complete list of symbols just before the index at the end of this book. In the Book

$$\alpha = 1/\varepsilon_0$$

where ε_0 is the permittivity of free space. As far as possible in this book ε_0 is used, for clarity, but α does slip in. So α is not used for the fine structure constant which instead is denoted by FSC. This is because α is introduced as an unknown constant in Chapter 1 of the Book before its fundamental role as $1/\varepsilon_0$ is clear, and the fine structure constant is not encountered until Chapter 3. In the parts of the book dealing with quantum mechanics we use $\exp(i\omega t - ikz)$ instead of the conventional $\exp(ikz - i\omega t)$ and this results in a change in sign of some operators and other expressions. This choice has been made because it is more natural to use $\exp i\omega t$ in the description of a stationary particle than $\exp(-i\omega t)$. SI units are used throughout. Vectors are in bold type, thus **p**. Convolution is denoted by $*$.

Chapter 1

A connection between electricity and gravity

1.1 Introduction

There is general agreement that physics is in need of a unifying fundamental theory which underpins the established but separate strands and reveals the status of the more uncertain areas. It should be based on as few assumptions as possible and on the minimum of fundamental constants. Our aim is to account for the whole of physics by starting with fundamental postulates and then developing a theory from them. This chapter introduces our postulates and it is shown from them how a connection can be made between electricity and gravity.

As the development proceeds, various entities will be introduced as a consequence of that development. This means that we cannot assume that we understand what a concept is until it has been introduced. This applies to all the familiar concepts of physics. For example, we cannot assume that we know what energy is before we start the development of the theory. Energy has to arise as a consequence of the theory. Also the physical significance of each entity needs to be established within the theory by understanding its interpretation in the world of observation. For example, at some point we shall introduce the electric field. However we cannot interpret this as the force per unit charge until we have introduced the concept of force and understood the nature of the forces between charged particles.

1.2 The postulates

The new theory is based on two postulates:

> **The First Postulate.** Each observer observes a three dimensional space filled by two continuous single-valued velocity fields which specify velocity vectors which are functions of time at each point in the three dimensional space.

A connection between electricity and gravity

When we talk about a field we are referring to an entity which is defined at each point in space. Velocity refers to the speed of something in a particular direction. So a velocity field refers to the speed in some direction of something at each point in space. Each one of the observers will consider himself or herself to be at rest and that some or all other observers are moving. No matter, this second postulate states that all observers will observe the same the type of thing.

> **The Second Postulate.** There is a special velocity magnitude such that, when one of the velocity field vectors at a selected point is observed by one observer to have the special magnitude, all observers observe that at the selected point the velocity field has the special magnitude.

The magnitude of the velocity is the speed. This second postulate is similar to Einstein's special relativity principle which states that the laws of physics are the same in all frames of reference moving at uniform speeds with respect to each other. One specific piece of physics concerns the speed of light. It follows from Einstein's special relativity principle that the speed of light will be measured to be the same in all frames of reference. However our postulate is not identical to this conclusion; we are just saying that there is a special speed, and if something is travelling at this speed in one frame of reference, it will be measured to be travelling at the same speed in all moving frames of reference. It is a matter in the development of the theory as to whether light travels at this special speed or not - we deal with this in Chapter 3. However a consequence of the similarity with the Einstein principle is that special relativity supplies a ready-made mathematical analysis and the new theory exploits the results of relativity mathematics extensively. We do not expect the reader to be familiar with this mathematics and we shall call attention to the few results that we need to use as they arise.

We are going to demonstrate that the First Postulate leads to there being two types of entity in the universe and we shall call them electric charge and gravitational mass. Our first step towards this goal is to show that there exists an entity extending throughout all space which we are free to call the gravitational mass density.

1.3 The Field Equations

Let's confine attention initially to one of the velocity fields. The postulates refer to a three dimensional space and to points within that space. The postulates do not specify what entity is moving at each point. Consider therefore a stationary point chosen by an observer. Now consider another point which is initially coincident with the chosen stationary point and which is moving with a velocity specified by the local velocity field. The moving point will reach successively the positions of other stationary points and we shall allow its velocity to take on the speed and direction specified by the local velocity field at these points. So the moving point is observed to traverse a trajectory in space.

Now consider a second point which traverses a second trajectory. Because the velocity field must remain continuous and single-valued, the two trajectories will map out flow lines which do not intersect. If we extend this to many moving points, the number of moving points is conserved, that is, points are neither created or destroyed. Since this is the case for every frame of reference, the conservation of the number of moving points is observed to be the case by all observers.

Now consider an observer who denotes position by x, y and z with respect to Cartesian, orthogonal, axes. The observer measures time, denoted by t, using a clock at rest and so is able to measure velocity in terms of change in position with time. The velocity field observed by the observer will be denoted by $\mathbf{u}(x,y,z,t)$ or simply $\mathbf{u}$. $\mathbf{u}$ is in bold type because it is a vector. It has a magnitude, i.e. speed, and it is in a particular direction which at the moment remains unspecified. Now let's increase the number of points so that we have a density of m points per unit volume. The flow of points, in units of points per unit area per unit time, denoted by $\mathbf{p}$, is given by $\mathbf{p} = m\mathbf{u}$. The magnitude of $\mathbf{p}$ we will write as p. So we have established that an entity m can exist and it is associated with a flow $\mathbf{p}$. We shall name the entity described by m as the gravitational mass density and $\mathbf{p}$ as the gravitational momentum density. We comment on these names further below.

We now turn attention to the second postulate. The special speed will be denoted by c. As remarked above the second postulate is similar to Einstein's special relativity principle. However c is a special speed by virtue of a point moving at c having the same speed c when

A connection between electricity and gravity

observed by different moving observers. As we explained, the physics here is not the same as in special relativity, but we make use of the same mathematics. Consider the following equation,

$$\frac{\partial^2 m}{\partial x^2} + \frac{\partial^2 m}{\partial y^2} + \frac{\partial^2 m}{\partial z^2} - \frac{1}{c^2}\frac{\partial^2 m}{\partial t^2} = -\frac{\rho}{\varepsilon_0}$$

(1.1)

It contains partial derivatives, and we shall refer to this type of equation a number of times in this chapter – but no matter; all we need to know is that the left hand side is in the form of a 'differential operator' operating on m and it allows us to find a new entity ρ on the right hand side. It is a consequence of special relativity mathematics that the same form of left hand side is obtained in all other moving frames of reference. This means that ρ on the right hand side can be obtained from m in any frame of reference by using this equation. We shall call ρ the electric charge density, and we are going to measure it in units of coulombs per cubic metre. The details of how we set up a system of measurements can only be tackled after the major elements of the theory are in place. Nevertheless we declare at the outset that we shall use SI units and we will introduce each measurement unit when appropriate. Thus charge is to be measured in coulombs. ε_0 is a constant. We are going to call it the permittivity of free space. You might have thought that its value could be arbitrary, i.e. just think of a value and stick to this value. This is exactly what is done in setting up the SI system of measurement, and we give some further details in Chapter 8. You might also ask what value we are to give to c. Is this also arbitrary? Well, yes it is and again the SI system adopts an arbitrary value, discussed in Chapter 8.

Since we can generate a new entity from m, let's do the same with **p**. Consider

$$\frac{\partial^2 \mathbf{p}}{\partial x^2} + \frac{\partial^2 \mathbf{p}}{\partial y^2} + \frac{\partial^2 \mathbf{p}}{\partial z^2} - \frac{1}{c^2}\frac{\partial^2 \mathbf{p}}{\partial t^2} = -\frac{\mathbf{j}}{\varepsilon_0}$$

(1.2)

This means that a new entity **j** can be obtained from **p** in any frame of reference using this equation. We shall call **j** the current density and is to be measured in amps per square meter. If we put

$$\mathbf{j} = \rho \mathbf{v}$$

1.3 The Field Equations

then since we know **j** and ρ then we can obtain **v**. It can be proved that charge is conserved and so **v** is a continuous and single-valued velocity field. The First Postulate requires that there are two velocity fields. The first is **u** above and we have found a second velocity field **v**. However we have not satisfied the Postulate yet because the differential operator could be applied to ρ and **j** to obtain new entities connected by a third velocity field. Indeed there could be an infinite sequence of new entities generated by successive applications of the operator and hence an infinite number of velocity fields. However the First Postulate states that there can only be two velocity fields and hence only two entities, m and ρ. We can arrange for this to be the case by linking ρ back to m by putting

$$\frac{\partial^2 \rho}{\partial x^2} + \frac{\partial^2 \rho}{\partial y^2} + \frac{\partial^2 \rho}{\partial z^2} - \frac{1}{c^2}\frac{\partial^2 \rho}{\partial t^2} = \frac{m}{\varepsilon_0}$$
(1.3)

so that applying the operator to ρ results in the first field m multiplied by a constant $1/\varepsilon_0$ on the right hand side. Comparing this equation with (1.1) above, the minus sign on the right hand side has disappeared. This change is required because a piece of mathematical manipulation shows that the system otherwise will reduce to a single entity. There is a parallel equation for obtaining **p** from **j**,

$$\frac{\partial^2 \mathbf{j}}{\partial x^2} + \frac{\partial^2 \mathbf{j}}{\partial y^2} + \frac{\partial^2 \mathbf{j}}{\partial z^2} - \frac{1}{c^2}\frac{\partial^2 \mathbf{j}}{\partial t^2} = \frac{\mathbf{p}}{\varepsilon_0}$$
(1.4)

So we have constructed a scheme which is consistent with the two postulates and contains two types of entity which we call gravitational mass and electric charge. Equations (1.1) to (1.4) are the Field Equations, a simultaneous set of partial differential equations which connect the electric and gravitational entities. It is the solutions of these equations which lead to all of the predictions of this new theory.

Comparing (1.3) with (1.1), then we can see that we can measure the gravitational mass density, m, in coulombs per cubic metre. The gravitational mass that we have introduced is a new concept. We could have called it gravitational charge, since it has units of coulombs, and we could have called **p** the gravitational current density instead of the gravitational momentum density. However charge is very familiar as

A connection between electricity and gravity

the electric charge and we shall show in Chapter 4 that gravitational mass in coulombs is proportional to inertial mass in kilograms, and gravitational momentum is proportional to inertial momentum. Hence our choice of names is to use gravitational mass and gravitational momentum.

From special relativity mathematics p and m are related by

$$m^2 - p^2/c^2 = m_{00}^2$$

(1.5)

where m_{00} is the gravitational mass density when it is at rest with respect to an observer. Analogously,

$$\rho^2 - j^2/c^2 = \rho_{00}^2$$

(1.6)

where ρ_{00} is the rest charge density. These results will be termed the relativistic constraints. It turns out that we do not have to investigate m_{00} and ρ_{00} further because they are discarded in Chapter 2.

1.4 The Field Equations and the development of the new theory

We have established that the two postulates lead to the existence of m and ρ. They are connected by the Field Equations (1.1) and (1.3). We have called them the gravitational mass density and the charge density respectively, but we could have called them by any pair of arbitrary names. We have chosen the names because, of course, eventually, the development of the theory will make clear they indeed correspond to what we conceptually call charge and gravitational mass densities. The development of the theory proceeds by seeking solutions of the Field Equations (1.1) to (1.4).

The journey in the chapters ahead takes us to the derivation of Coulomb's law of electrostatic attraction and repulsion, and to Newton's law of gravitation. ε_0, the permittivity of free space, will play its role in Coulomb's law in its conventional role. However, inspection of (1.1) shows it has dimensions of L^2 where L stands for the dimensions of length. This also means that the 'Farad' in conventional electromagnetic theory turns out to have dimensions of L^3, but this is entirely permissible, although unconventional. For the interaction between gravitational masses there will be a form of Coulomb's law. Comparison of the form of Coulomb's law and Newton's law of gravitation allows us to make a theoretical prediction of Newton's gravitational constant, but these results lie a long way ahead.

1.5 Summary

This chapter has set out the two postulates on which the new theory is based. It is shown that entities that we choose to call gravitational mass and electric charge follow from the postulates. Electricity and gravity are connected in that the various electrical and gravitational entities are linked by a set of Field Equations. The development of the new theory proceeds by finding solutions of the Field Equations. The various concepts of physics are being introduced as mathematical entities, but it is a long road ahead in the following chapters before we reach their familiar physical roles.

Chapter 2

A model for fundamental particles

2.1 Introduction

Experimental physics has established that matter (solids, liquids, gases) on Earth and elsewhere is composed of atoms. Atoms in turn are composed of electrons and nuclei; nuclei are composed of protons and neutrons. There are other fundamental particles which include muons, pions, neutrinos and so on. The current accepted view of the nature of fundamental particles is captured in the standard model. Particles are classified into leptons and hadrons, and hadrons are further divided into baryons and mesons. The composition of hadrons is further analysed based upon their quark content. Thus the standard model reduces the collection of fundamental particles down to leptons (electrons and electron neutrinos, muons and muon neutrinos, and the tau and tau neutrinos), and three pairs of quarks (down, up; strange, charm; bottom, top). Particles are also characterised by their spin which is related to their internal angular momentum. It is the spin parameter through which a connection is made in this chapter between the new theory and the measured properties of particles and therefore with this aspect of the standard model. The development of models for protons, neutrons, pions, kaons, electrons, muons and neutrinos is tackled in Chapters 6, 7 and 14. It is in Chapter 15 that the connection is made in the new theory with the quark content of hadrons.

The previous chapter introduces the Field Equations which govern the behaviour of the gravitational mass and charge densities and the associated gravitational momentum and current densities. The problem tackled in the present chapter is how to develop solutions of the Field Equations in order to account for the existence and properties of fundamental particles. In particular, we are going to account for the structure of particles.

2.3 Particle internal fields

2.2 The particle model

The particle model is based on the following concepts.
Particles are not points and therefore they have structure. They are objects with a volume contained within a boundary.

The internal densities are composed of oscillatory and steady state components. It will be seen in Chapter 4 that the angular frequency of the oscillation gives rise to the inertial mass of the particle. The steady state charge density drops to near zero at the boundary. There may be a transition region over which this reduction occurs but in this chapter the steady state charge density is taken to be discontinuous at the boundary. Gravitational mass density and electric charge density vary with position within the particle and their integrals are the particle's gravitational mass and electric charge respectively. The oscillatory densities extend beyond the boundary into the space beyond.

Oscillatory waveforms cannot exist in isolation from steady state levels. This is because gravitational mass or charge densities are not to become zero – this is elaborated on in section 2.3 below. This allows a connection to be made between the steady state levels and the oscillatory amplitudes.

The particle is a source of external oscillatory waveforms. Where the oscillatory waveforms encounter other particles, referred to as target particles, the oscillatory waveforms require steady state values to be associated with them, and this mechanism gives rise to steady state components in the target particle due to the distant source particle. A very important conclusion is that the charge and gravitational mass required within target particles due to the source particle comes from the charge and gravitational mass already in the target particles.

2.3 Particle internal fields

We turn our attention to the internal fields for the particle at rest, the gravitational mass density m, the charge density ρ, the gravitational momentum density **p** and the current density **j**. We can construct solutions to the Field Equations in the form of sums of oscillatory and steady state components. We write the oscillatory component as $m_A(\cos \omega_0 t + i \sin \omega_0 t)$ where i is the square root of -1. It is a standard result that this expression is equal to $m_A \exp(i\omega_0 t)$ and we put

A model for fundamental particles

$$m = m_A \exp(i\omega_0 t) + m_B$$

and similarly

$$\rho = \rho_A \exp(i\omega_0 t) + \rho_B$$

where ω_0 is the particle's oscillation angular frequency, m_A and ρ_A are the oscillatory component amplitudes and m_B and ρ_B are the steady state components.

Let's consider the steady state component. Given m_B and ρ_B there is an immediate solution for the steady state gravitational momentum and current densities,

$$p_B = cm_B \qquad j_B = c\rho_B$$

where the motion is circular about the z axis within the particle, see Figure 2.1.

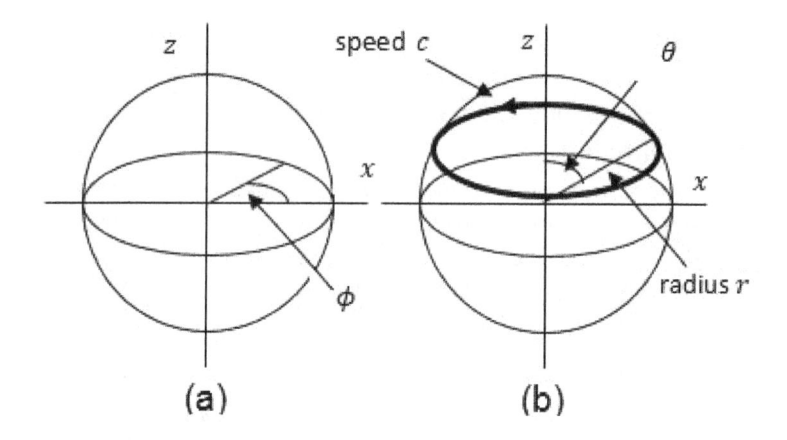

Figure 2.1 Spherical co-ordinates use the parameters r, θ, ϕ. In Figure (a) ϕ is the angle about the z axis from the x axis. In Figure (b) r is the radius and θ is the angle between the radial line and the z axis. The thickened circle shows an example of the circulation of charge or gravitational mass with speed c around the z axis keeping θ constant

This solution ensures that the amount of gravitational mass and charge within the particle remain constant.

We now turn attention to the relativistic constraints introduced in the last chapter. We need to take both the steady state and oscillatory components into account. However if we take the amplitude of the

2.3 Particle internal fields

oscillatory components to be small compared to their respective steady state levels, and this is discussed in the Book, then the relativistic constraints, equations (1.5) and (1.6), simplify to

$$m_B^2 - p_B^2/c^2 = 0 \quad \text{and} \quad \rho_B^2 - j_B^2/c^2 = 0$$

and so the rest values m_{00} and ρ_{00} introduced in Chapter 1 become zero and can now be discarded. However there is the requirement that m and ρ do not ever become zero because otherwise m and ρ would be allowed to reduce towards m_{00} and ρ_{00} respectively and the speed will drop below c. This means that the oscillatory components cannot occur in isolation – they need to be superimposed on a steady state level in order to avoid the gravitational mass and charge densities passing through zero. This condition can be used to estimate the maximum allowable values of m_A and ρ_A by simply putting $m_A = m_B$ and $\rho_A = \rho_B$. These estimates are used in Chapter 6 in connection with the proton model. A consequence of the oscillatory components not being unaccompanied allows a connection to be made between steady state and oscillatory values. We require that $m_A < m_B$ and $\rho_A < \rho_B$ inside a stable particle and this can be ensured if m_B is a function of m_A.

We can obtain this important relationship between the oscillatory amplitudes and the steady state levels as follows. We need to introduce the idea that the particle is connected to gravitational mass and charge densities outside its boundary. Let's denote its external oscillatory densities by m_{AE} and ρ_{AE} which are associated with steady state densities m_{BE} and ρ_{BE} respectively. Consider a point far from any particles. Suppose that the steady state gravitational mass density in the far field of our chosen particle is connected to the oscillatory amplitude by

$$m_{BE} = A_m m_{AE} m_{AE}^* \tag{2.1}$$

where A_m is a constant and m_{AE}^* is the complex conjugate of m_{AE}. Suppose that other particles also contribute and that their oscillatory components are at the same angular frequency and their amplitude contributions happen also to be m_{AE} but randomly distributed in phase. The total steady state gravitational mass density is $m_{BT} = n A_m m_{AE} m_{AE}^*$ where n is the number of contributing particles. If the resultant oscillatory amplitude is m_{AT} we require that

A model for fundamental particles

$$m_{BT} = A_m m_{AT} m_{AT}^*$$

This means that we require that $m_{AT} = n^{1/2} m_{AE}$ which is the case when the phases of the individual oscillatory components are uncorrelated and are distributed randomly over 2π. Thus we have found a relationship (2.1) that is sufficient to deal with this particular summation of steady state densities in the far field. Since there is continuity of the oscillatory densities and steady state potentials back to the source particles, then the relationship (2.1) applies inside particles.

We could have negative steady state densities and we require in this case that

$$m_B = -A_m m_A m_A^*$$

where A_m remains a positive constant. By the same arguments using ρ_{AE} and ρ_{BE} we can write

$$\rho_B = \pm A_\rho \rho_A \rho_A^*$$

Thus we have shown that the steady state densities are proportional to the corresponding oscillatory amplitudes squared. These results are used extensively in the development of the theory in the following chapters. We refer to A_m and A_ρ as the steady state constants and they are fundamental constants. These constants are not part of conventional physics and we take steps in Chapter 8 to obtain their numerical values. With m_{00} and ρ_{00} zero, m_{00} and ρ_{00} can be dropped from the set of fundamental constants. Instead the steady state constants A_m and A_ρ are included and the set now includes c, ϵ_0, A_m and A_ρ. We introduce the constant $\hbar$ to complete the set of five fundamental constants in Chapter 4.

The steady state gravitational mass contained within the particle boundary is given by

$$M_0 = \int_V m_B \, dV$$

where V is the volume within the particle boundary. M_0 is referred to as the particle gravitational mass. The steady state electric charge contained within the particle boundary, referred to as the particle electric charge, is given by

$$q = \int_V \rho_B \, dV$$

Although the sum of steady state and oscillatory gravitational mass densities or the sum of steady state and oscillatory charge densities must not pass through zero, this does not preclude the steady state values being negative, as discussed above. Consider a solution set of the oscillatory and steady state Field Equations with A_ρ and A_m positive. Now consider a new particle for which $A_\rho \to -A_\rho$ and $A_m \to -A_m$. The oscillatory components within the particle are unchanged and the oscillatory and steady state Field Equations continue to be satisfied. The particle's gravitational mass and electric charge become negative. This mechanism can account for anti-particles. There is the possibility that there could be other solutions, for example with $A_\rho \to -A_\rho$ with A_m unchanged. This gives an alternative mechanism for anti-particles and they will have positive gravitational mass. These possibilities are discussed further in Chapters 6 and 11.

2.4 Particle categories

The general problem for determining particle internal solutions depends on finding solutions of the oscillatory and steady state scalar Field Equations. Using spherical co-ordinates, see Figure 2.1, it is shown in the Book that we can obtain solutions of the form

$$m_A = R_m(r)\Theta_m(\theta)\Phi_m(\phi) \qquad \rho_A = R_\rho(r)\Theta_\rho(\theta)\Phi_\rho(\phi)$$

where

$$\Phi_m = \Phi_{m0} \exp(\pm is\phi) \qquad \Phi_\rho = \Phi_{\rho 0} \exp(\pm is\phi)$$

where Φ_{m0} and $\Phi_{\rho 0}$ are constants. s is a constant (not necessarily an integer) which we call the spin quantum number because later in Chapter 4 it turns out to be related to a particle's internal inertial angular momentum – but we have not yet introduced the concept of inertial angular momentum. The following scheme of particle categories is based on the value of s – the details are in the Book:

Group A (mesons), $s = 0$
Group B (baryons, leptons), $s = 1/2$

A model for fundamental particles

Group C (mesons), $s = 1$
Group D (baryons), $s = 3/2$

These are referred to as single omega solutions because the angular frequency is the same throughout the particle. There are more ways of constructing particles which involve a mix of solutions with differing angular frequencies within a single particle. The discussion of these matters has to wait until Chapter 14.

This chapter has given some general features of the solutions of the Field Equations. Features specific to particular particles are given later in Chapters 6 and 7 in which models for the proton, neutron, electron, electron neutrino, muon, pion and kaon are proposed. These proposals provide amplification of the allocation of mesons to groups A and C, leptons to group B and baryons to groups B and D. It is shown in Chapter 8 that the values of a number of particle parameters are derivable from the five fundamental constants with no additional constants required.

2.5 Particle fields for a spherical particle

We now consider the consequences of the case where the particle charge magnitude is much greater than the gravitational mass magnitude. This requires that

$$|\rho_B| \gg |m_B|$$

When the various parameters are quantified in Chapter 8 it will be seen that the predicted particle charge is around twenty orders of magnitude greater than the predicted gravitational mass. In what follows we shall assume for convenience that the electric charge density ρ_B is positive.

It is useful to consider a notional spherically symmetric charged particle bounded by an inner spherical surface, radius r_1, and an outer spherical surface, radius r_2. We are going to write down some more differential equations, but only in order to point out basic features of some of the solutions. The internal steady state charge density, ρ_B, satisfies the Field Equation

$$\frac{d^2\rho_B}{dr^2} + \frac{2}{r}\frac{d\rho_B}{dr} = \frac{m_B}{\varepsilon_0}$$

where we are now using spherical coordinates and r is the radius from

2.5 Particle fields for a spherical particle

the centre of the co-ordinates. It is in a simplified form of Poisson's equation, connecting ρ_B on the left hand side to m_B on the right hand side. There are two features of the solution to this equation we need to consider. The first feature is that if the right hand side is zero, we obtain a simplified form of Laplace's equation. This can arise where the right hand side is very small in comparison with the left hand side and we are justified in neglecting the right hand side. The solution to this equation is in the form of

$$\rho_B = B/r$$

where B is a constant. The second feature is that we can always add a B/r term to the solution of the Poisson's equation. We now exploit these two features in our particle model. We put

$$\rho_B = \rho_{B1} + \rho_{B2}$$

We explain ρ_{B2} shortly. ρ_{B1} satisfies

$$\frac{d^2 \rho_{B1}}{dr^2} + \frac{2}{r}\frac{d\rho_{B1}}{dr} = 0$$

This is in the form of Laplace's equation with a zero on the right hand side, and so ρ_{B1} internally can be proportional to $1/r$, see Figure 2.2.

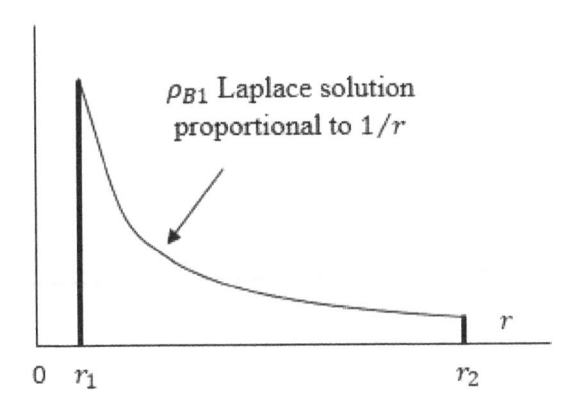

Figure 2.2 The particle internal charge density ρ_{B1} versus radial distance r. There are discontinuities at the particle inner radius r_1 and at the particle outer radius r_2

A model for fundamental particles

A B/r variation allows the charge density to be arbitrarily large and this allows the electric charge to be much greater than the gravitational mass. It is proposed that ρ_{B1} is bounded by discontinuities at r_1 and r_2, i.e. it rises rapidly at r_1 and decays rapidly to zero at r_2 at the particle boundary, as shown in Figure 2.2. The justification for the discontinuities is discussed in Appendix A where there is more detail for the specialists.

Now let's examine the variation of m_B. We have

$$\frac{d^2 m_B}{dr^2} + \frac{2}{r}\frac{dm_B}{dr} = -\frac{\rho_B}{\varepsilon_0}$$

(2.2)

and because internally $\rho_B \gg m_B$ we are not able to make the simplification by putting the right hand side to zero and we show this solution as a Poisson solution on Figure 2.3. We can always add internally a Laplace component to the Poisson solution for m_B, with a discontinuity at the boundary. An important consideration is that the oscillatory amplitude does not exceed the steady state level, discussed above in Section 2.3. There is a minimum value of ρ_B at the outer boundary to ensure that this is the case. Putting $\rho_A = \rho_B$ leads to $\rho_B = 1/A_\rho$ and so we require $\rho_B = 1/A_\rho$ at the boundary. This is dealt with in Chapter 8.

We now account for ρ_{B2} by putting

$$\frac{d^2 \rho_{B2}}{dr^2} + \frac{2}{r}\frac{d\rho_{B2}}{dr} = \frac{m_B}{\varepsilon_0}$$

(2.3)

This is a Poisson equation and just as m_B is very much less than ρ_{B1} then ρ_{B2} is very much less than m_B and is sketched on Figure 2.3. It is straightforward to show from the steady state Field Equations that at a spherical particle's surface and beyond

$$m_{BE} = \frac{q}{4\pi\epsilon_0 r}$$

(2.4)

$$\rho_{B2E} = -\frac{M_0}{4\pi\epsilon_0 r}$$

(2.5)

2.5 Particle fields for a spherical particle

where m_{BE} is related to the oscillatory amplitude m_{AE} by

$$m_{BE} = A_m m_{AE} m_{AE}^* \qquad (2.6)$$

and ρ_{B2E} is related to ρ_{A2E} by

$$\rho_{B2E} = A_\rho \rho_{A2E} \rho_{A2E}^* \qquad (2.7)$$

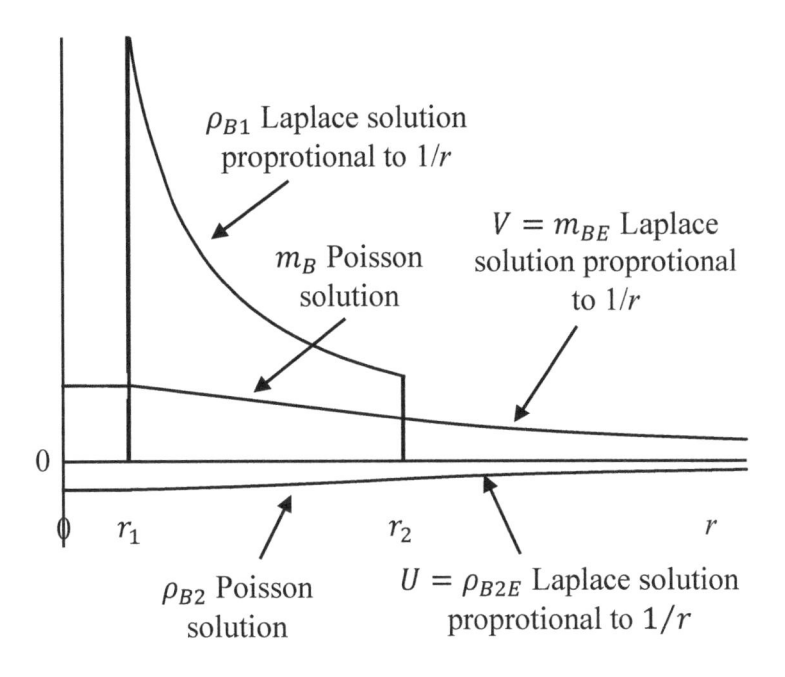

Figure 2.3 The particle internal and external steady state charge and gravitational mass densities as functions of the radial distance r

Thus we can view our particle as a source of oscillatory fields m_{AE} and ρ_{A2E} which require associated steady state fields m_{BE} and ρ_{B2E}. These latter are related by

$$\frac{m_{BE}}{\rho_{B2E}} = -\frac{q}{M_0}$$

Since q is very much greater than M_0 then $|\rho_{B2E}|$ is very much less than m_{BE}. Since ρ_B is very much greater than m_B then ρ_{B1} is very much greater than m_B. We can now complete the description of Figure 2.3. If another particle, which we call a target particle, were placed at successive positions beyond the boundary, the external fields due to the source particle appearing within the target particle are the oscillatory gravitational mass and charge densities, m_{AE} and ρ_{A2E}, together with their associated steady state fields, m_{BE} and ρ_{B2E}.

Hence we extend m_B as m_{BE} and proportional to $1/r$ into the space surrounding the particle, and similarly we extend ρ_{B2} as ρ_{B2E} which is also proportional to $1/r$. m_{BE} is continuous with the internal m_B when m_B is only a Poisson solution. If we have added an internal Laplace solution to m_B there will be a discontinuity at the boundary. This will be the case when the particle is negatively charged but has positive gravitational mass.

2.6 The effect of the source particle external steady state densities on target particles

We have written down the external steady state densities from the source particle, equations (2.4) and (2.5), assuming that the source particle charge and gravitational mass are unaffected by the presence of other particles and are therefore conserved. We now examine in more detail the effect of the source particle external oscillatory densities on target particles. As already stated above, the overlap of the external oscillatory densities with a target particle results in steady state densities appearing in the target particles given by (2.6) and (2.7).

Charge and gravitational mass need to be conserved. This must be the case for an arbitrary number of target particles at arbitrary distances from a source particle, taking into account that each source particle is also a target particle and each target particle is a source particle. For the gravitational mass and charge to be conserved, it is sufficient that the gravitational mass and charge required for the steady state densities from the source particle induced inside the target particle come from the gravitational mass and charge already in the target particle. This principle is so important and fundamental to the derivation of classical mechanics and quantum mechanics in Chapter 4, that we repeat it:

2.6 The effect of the source particle external steady state densities on target particles

The steady state gravitational mass and charge induced in a target particle by the external oscillatory densities from source particles are contained within the gravitational mass and charge already in the target particle.

This principle ensures that both charge and gravitational mass of a target particle are unchanged from their original values. This can be expressed as follows, assuming for simplicity that all fields are uniform over the particle volume and where m_{BE} and ρ_{B2E} are derived from the source particle external oscillatory densities using (2.6) and (2.7),

$$m_{BE} + m_B = m_{B0} \tag{2.8}$$

$$\rho_{B2E} + \rho_B = \rho_{B0} \tag{2.9}$$

where m_B and m_{B0} are the target particle's present and original steady state gravitational mass densities and ρ_B and ρ_{B0} are the present and original steady state charge densities. This means that the steady state densities in the target particle, m_B and ρ_B, which derive from the target particle oscillatory densities must adjust so as to satisfy (2.8) and (2.9). These concepts may prove difficult to accept. Let's pose some questions and provide answers.

(1) The external densities from massive and charged objects permeate into any surrounding particles. Are charged and massive objects surrounded by extra gravitational mass and charge?

Equations (2.8) and (2.9) ensure that the source particle external oscillatory densities do not generate more charge and gravitational mass than already there due to the target particles. Hence there is no extra steady state gravitational mass density in the vicinity of a charged object and there is no extra steady state charge density in the vicinity of a massive object.

(2) The external steady state gravitational mass density due to a negatively charged object is negative. Does this imply the existence of negative gravitational mass?

A model for fundamental particles

The source particle induces negative gravitational mass density in the target particle, but the target particle gravitational mass remains constant, i.e. m_B adjusts as required by (2.8). There is the possibility that particles could have negative gravitational mass – this is mentioned in Section 2.3 and is discussed in Chapters 6 and 11.

(3) How can there be a charge density induced in a neutral target particle?

If a neutral particle is composed of separate positively charged and negatively charged sections, then equation (2.9) can be applied to each section separately. It will be seen that neutral particles examined in later chapters are composed in this manner.

(4) How can there be a charge density induced by a neutral source particle in a target particle?

Since neutral particles are composed of positively and negatively charged sections, each section will give rise to its own ρ_{B2E} type external density in target particles.

2.7 The external potentials

The right hand side of equation (2.4) is in the familiar form of the classical electromagnetism expression for the electric potential in a charge and current free region. Let's return to equation (2.2). Put $V = m_B$ and we obtain

$$\frac{d^2V}{dr^2} + \frac{2}{r}\frac{dV}{dr} = -\frac{\rho_B}{\varepsilon_0}$$

which is the electromagnetic result connecting the electric potential V and the charge density ρ_B. So we can interpret the external steady state gravitational mass density as giving rise to the external electric potential. This means that (2.4) can be recast as

$$m_{BE} = V = \frac{q}{4\pi\varepsilon_0 r}$$

to provide the familiar expression for the external electric potential assuming that the source particle is the only source present affecting

2.8 The background

the target particle and forgetting about all other source particles in the universe. Similarly we can put $U = \rho_{B2}$ and obtain from equation (2.3)

$$\frac{d^2U}{dr^2} + \frac{2}{r}\frac{dU}{dr} = \frac{m_B}{\varepsilon_0}$$

which means that (2.4) can be recast as

$$\rho_{B2E} = U = -\frac{M_0}{4\pi\epsilon_0 r}$$

and this is the expected form for the external gravitational potential in mass free space. We indicate the connections between V, U and m_{BE}, ρ_{B2E} on Figure 2.3. We can recast (2.8) and (2.9) for the target particle as

$$V + m_B = m_{B0} \tag{2.10}$$

$$U + \rho_B = \rho_{B0} \tag{2.11}$$

The formalism has yet to attach physical significance to V and U. This is not done until Chapter 4 where the familiar concept that V is the potential energy per unit charge emerges. However we note that if we multiply the left hand sides of (2.10) and (2.11) together and equate to the product of the right hand sides, and then multiply both sides by the particle volume, we obtain, neglecting the term in UV,

$$(m_B \rho_B - m_{B0}\rho_{B0})\mathcal{V} + qV + M_0 U = 0$$

and in Chapter 4 this becomes the equation of energy.

2.8 The background

The external oscillatory densities are provided by the source particles. In between the target particles there must be some mechanism giving rise to steady state densities on which the source particle external oscillatory densities can be superimposed. This mechanism has to conserve gravitational mass and charge. We also need a mechanism for continuity of the signs of A_m and A_ρ so that the correct signs occur

A model for fundamental particles

when inducing steady state densities in target particles. If we had particles everywhere we would solve these problems. However they cannot be as prominent as the target particles we have already introduced, because we would then call them target particles. It is proposed that there is a background of particles, but very thinly distributed, and Equations (2.10) and (2.11) apply but with very much reduced values of $V, U, m_B, m_{B0}, \rho_B$ and ρ_{B0}, all reduced by the same factor. The factor is evaluated in Chapter 17. So free space is not filled with extra gravitational mass and charge induced by the external oscillatory densities from source particles.

2.9 Summary

We have constructed a fundamental particle model. We have internal solutions of the Field Equations, and external solutions. Each consists of oscillatory and steady state components. The dominant internal component is the steady state charge density, accounting for the particle's electric charge. Similarly the particle has gravitational mass. The oscillatory fields extend into the surrounding space and induce steady state fields within target particles. The external steady state solutions are the potentials and it is through these that any one particle interacts with other particles, explored in the next three chapters. The steady state gravitational mass and charge induced in target particles by the potentials from source particles are contained within the gravitational mass and charge already in the target particle. The particle structure used for illustration here is spherical, but others are based on a wide variety of structural types, as we shall see later when we propose structures for the proton, neutron, kaons, pions, the electron, electron neutrino and many more particles.

Chapter 3
Photons

3.1 Introduction

Chapter 2 has shown how the oscillatory gravitational mass and charge densities from a source particle extend into the surrounding space. These induce steady state densities in target particles, and these steady state densities behave as external electric and gravitational potentials. The purpose of the present chapter is to find the external oscillatory solutions, and hence to determine how potentials are transmitted into the space surrounding a source particle, and to explore the mechanism for signalling a change in potential. The oscillatory solutions found lead to radiation field quantisation and to a model for photons. The conclusions from this chapter are used in the next chapter on classical and quantum mechanics.

The first step is the solution of the Field Equations in the form of travelling waveforms. The solutions are in the form of sums of products of a sinusoidal dispersive carrier and a Hermite function. The Hermite functions satisfy a harmonic oscillator eigenvalue equation and as a consequence there is quantisation of the radiation field. Next the connection between the travelling solutions and external potentials is examined. A way is required for a particle to signal to the surrounding space that it is changing its angular frequency and this leads to a model for the photon. It is established that an entity we call the 'energy' in photon waveforms is quantised, and that they also transport gravitational mass. A major conclusion is that charged particles are required to have a common value of charge magnitude in order to interact with a common set of photons.

3.2 Omega waveforms

The first step is to introduce the idea of an omega waveform. The omega waveforms gives rise to the electric and gravitational potentials as we shall see. We start with the oscillatory fields m_{AE} and ρ_{AE}, with

Photons

an angular frequency of ω, introduced in the previous chapter. From Chapters 1 and 2,

$$\frac{\partial^2 m_{AE}}{\partial x^2} + \frac{\partial^2 m_{AE}}{\partial y^2} + \frac{\partial^2 m_{AE}}{\partial z^2} - \frac{1}{c^2}\frac{\partial^2 m_{AE}}{\partial t^2} = -\frac{\rho_{AE}}{\varepsilon_0}$$

$$\frac{\partial^2 \rho_{AE}}{\partial x^2} + \frac{\partial^2 \rho_{AE}}{\partial y^2} + \frac{\partial^2 \rho_{AE}}{\partial z^2} - \frac{1}{c^2}\frac{\partial^2 \rho_{AE}}{\partial t^2} = \frac{m_{AE}}{\varepsilon_0}$$

Travelling wave solutions of these equations are now sought. We take the direction of travel to be along the z axis. It is shown in the Book that there exist solutions which are composed of components in the form of products of a Hermite type function of $z - ct$, a carrier described by a sinusoid, and a function of x, y,

nth component of $m_{AE} = m_{xyn} a_n h_n(z - ct) \exp(i\omega' t - ik_{mn}' z)$
$$(3.1)$$

nth component of $\rho_{AE} = \rho_{xyn} b_n h_n(z - ct) \exp(i\omega' t - ik_{pn}' z)$
$$(3.2)$$

where a_n and b_n are constants. Details regarding k_{mn}' and k_{pn}' are in Appendix B and the Book. The fields are quantised in that the $h_n(z'')$, where

$$z'' = \frac{\omega'}{c\sqrt{2n_0 + 1}}(z - ct)$$

are solutions of the Hermite differential equation and there are solutions when the parameter n is an integer. n_0 is a central value of n and we discuss that shortly. The solution components (3.1) and (3.2) apply when n is much greater than unity. n determines the number of zero crossings of the function and Figure 3.1 shows an example for $n = 10$. In the central region, around the origin, the Hermite function is sinusoidal, and it can be approximated by $\cos(\omega' t - k'z)$ where $k' = \omega'/c$. The product of this sinusoid and the carrier sinusoid results in a travelling wave of the form

$$A_v \cos(\omega t - kz)$$

for the gravitational mass density components where $\omega = 2\omega'$ and $k = 2k'$, and A_v is a constant amplitude. We explain v shortly. The

3.2 Omega waveforms

amplitude is B_ν for the charge density components. The Hermite function tends to zero away from the origin, as illustrated in Figure 3.1, and this means that the waveform is bounded by the extent of the Hermite function, and the effect is to limit the waveform to a wave packet travelling at the special speed c.

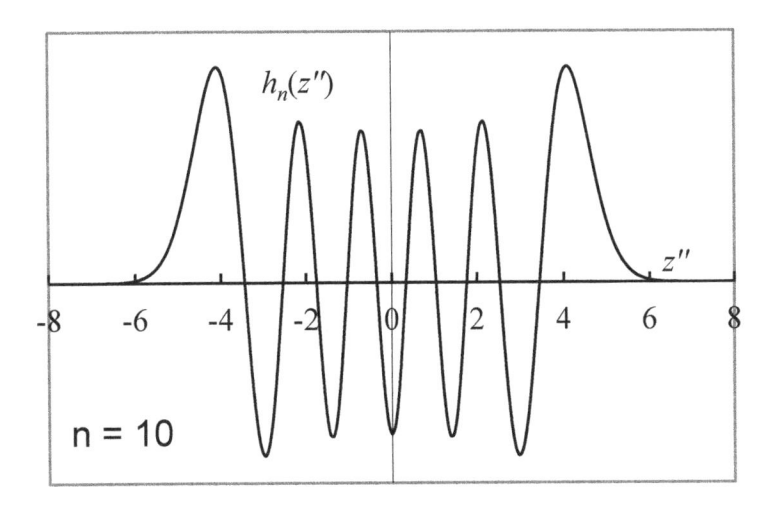

Figure 3.1 Example of a Hermite function illustrating the extent of sinusoidal behaviour when $n = 10$

We have not talked yet about the complete solution. It is obtained by taking a summation of waveforms of the type of equations (3.1) and (3.2) over n. The amplitudes fall away for n departing from a central value n_0, and so it is useful to introduce the integer ν where $\nu = n - n_0$. We can arrange that the gravitational mass density components only occupy odd values of ν and electric charge density components the even, or vice versa. Examples are shown in Figures 3.2 and 3.3 in which A_ν and B_ν are plotted versus ν. We define Type A as the category where n_0 (i.e. $\nu = 0$) is occupied by an m component, and Type B where n_0 is occupied by a ρ component. Our concern now is with Type A with gravitational mass density sinusoids only and from Figure 3.3 we can see that the algebraic sum of the charge density sinusoid amplitudes is zero.

Photons

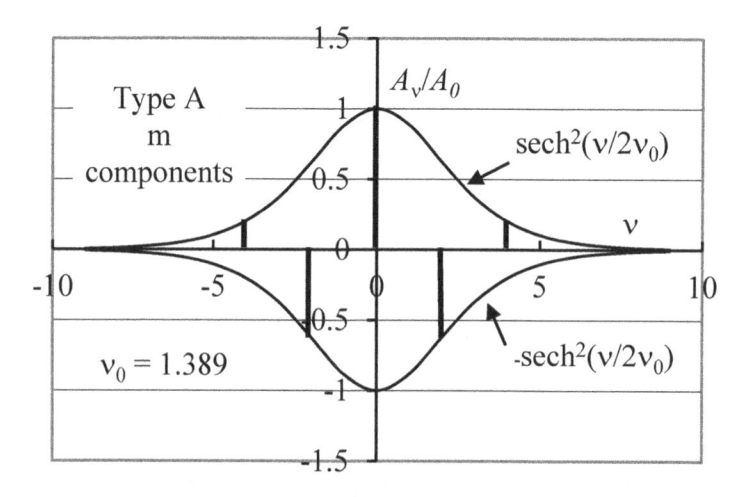

Figure 3.2 Example of Type A components for *m*, at ν = 0, ±2, ±4, etc.

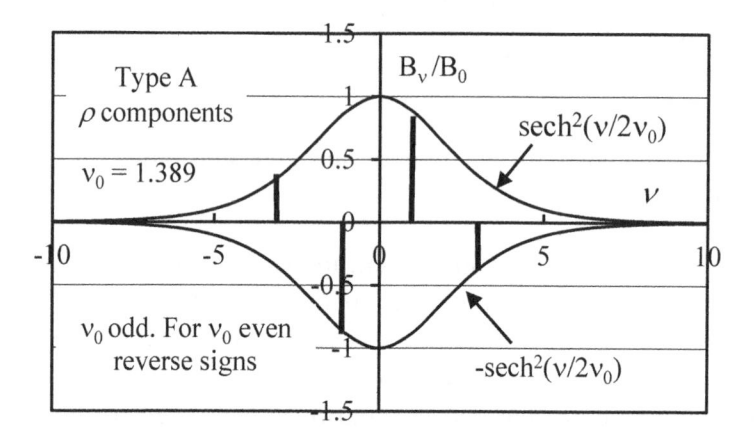

Figure 3.3 Example of Type A components for ρ for ν₀ odd, at ν = ±1, ±3, etc. For even ν₀ the signs multiplying the magnitudes are reversed

This means we can describe the situation where there is a non-zero electric potential and a zero gravitational potential. The algebraic sum of the amplitudes with respect to A_0 is given by

$$P \equiv \frac{\Sigma_\nu A_\nu}{A_0} = \sum_\nu \text{Sign}\left(\frac{S_\nu}{S_0}\right) \text{sech}^2\left(\frac{\nu}{2\nu_0}\right)$$

26

3.3 The connection between the external potentials and the omega waveforms

where Sign is a function which results in the ± behaviour shown in the Figures 3.2 and 3.3 and $\text{sech}^2(v/2v_0)$, where v_0 is a constant, describes the shape of the envelopes shown on Figures 3.2 and 3.3. P is plotted versus v_0 on Figure 3.4. P reduces as v_0 increases. This is in contrast to the behaviour of the 'energy' factor Q which increases with v_0 and which is introduced below.

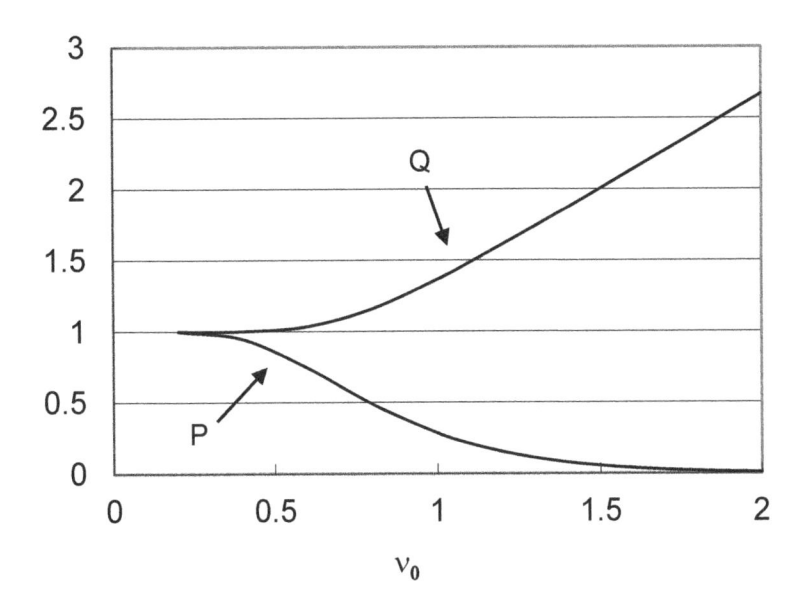

Figure 3.4 P and Q versus v_0

Now that we know how the a_n and b_n behave via the A_v and B_v, we have found the solutions for the oscillatory density fields m_{AE} and ρ_{AE} at angular frequency ω. We shall refer to these solutions as the omega waveforms.

3.3 The connection between the external potentials and the omega waveforms

We explore in this section the connection between the omega waveforms and the potentials originating from a source particle. A charged particle has external fields at radius r,

Photons

$$V = \frac{q}{4\pi\epsilon_0 r}$$

$$U = -\frac{M_0}{4\pi\epsilon_0 r}$$

where the particle charge is q and the gravitational mass is M_0. We focus on the electric potential but similar considerations apply to the gravitational potential. The first thing we need to note is that we require a travelling wave solution to the Field Equations in spherical coordinates. Instead of $\cos(\omega t - -kz)$ as discussed above, the solution for m_{AE} is proportional to the inverse of the distance from the source particle in the form,

$$m_{AE} = \frac{B}{r}\cos(\omega t - kz)$$

(3.3)

where B is a constant. Incoming and outgoing spherical travelling waves can be superimposed to give radial standing waves. This means that the electric potential due to the source particle can arise within a target particle by taking $m_{BE} = A_m m_{AE} m_{AE}^*$. However this leads to the following difficulty. The spherical oscillatory amplitude in (3.3) is proportional to $1/r$ and therefore the steady state level is proportional to $1/r^2$, whereas we require $m_{BE} = V$ to be proportional to $1/r$. This can be resolved as follows.

We propose the following mechanism for the propagation of the omega waveform – it takes place entirely within target particles external to the source particle. In Chapter 2, Section 2.8, we introduce the concept of the background, i.e. that in the space intervening between the source particle and a localised target particle, there is a background of particles, somehow distributed thinly (we propose a more detailed model in Chapter 17), so that the oscillatory waveform is accompanied by a steady state level throughout. The steady state level also transmits the sign of the potential. This background converts the $1/r$ variation for the oscillatory amplitude to $1/r^{1/2}$ as follows. A target particle, which itself oscillates at ω_T, will contain an oscillatory component at ω induced by the arrival of the omega waveform from the source particle. The particle, in the same way that it radiates the ω_T waveform, will also radiate components at ω. We can envisage a

3.3 The connection between the external potentials and the omega waveforms

spherical array of similar particles in the background, as sketched in Figure 3.5, interacting with an outgoing ω waveform.

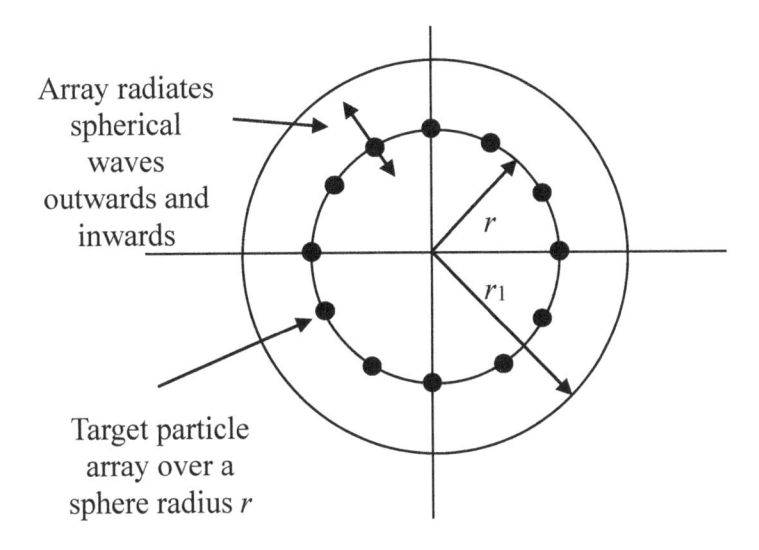

Figure 3.5 Background particle array re-radiating

The particles radiate both inwards and outwards. The inwards components from successive spherical arrays sum incoherently (or do they provide a backward wave? – see Appendix B, Section B.6). Suppose that we have achieved the amplitude variation equal to $r_0^{1/2}/r^{1/2}$ where r_0 is a constant, for the outgoing ω waveform at radius r and which impinges on the spherical array at r_1. The re-radiated amplitude outwards from the array is proportional to

$$\frac{1}{2}\left(\frac{r_0^{1/2}}{r^{1/2}}\right)\frac{1}{r_1}$$

at radius r_1 which adds to the outgoing amplitude. Summing over all spherical arrays out to radius r_1 results in an outgoing amplitude at r_1, where $(r_0^{1/2}/r^{1/2})dr$ is the gravitational mass per unit area as is B in (3.3),

$$\int_{r_0}^{r_1} \frac{1}{2}\left(\frac{r_0^{1/2}}{r^{1/2}}\right)\frac{1}{r_1} dr = \frac{r_0^{1/2}}{r_1^{1/2}}$$

for $r_1 \gg 0$ and which is what is required for consistency. Thus the steady state level is proportional to $1/r$ and so we can arrange that the resulting $m_{BE} = A_m m_{AE} m_{AE}^* = V$.

3.4 Photons

In this section we introduce a model for photons based on the requirement that source particles need to be able to signal a change in their angular frequency into the surrounding space. Five major properties are deduced, listed in section 3.5 below, which are used in the next chapter. First a digression is required to discuss further how the steady state component arises from the oscillatory omega waveforms. Consider for real amplitudes

$$2A_m |m_A|^2 \cos^2 \omega_0 t = A_m m_A m_A^* + A_m m_A m_A^* \cos 2\omega_0 t$$

This results in the steady state component $A_m m_A m_A^*$ provided that the $2\omega_0$ term is neglected. A way of describing this process is to say that it is square law detection followed by low frequency filtering. This construction can be applied to the case when a particle changes its state and the frequency changes from ω_1 to ω_2 as follows.

The source particle must signal somehow to the external world that it is now at angular frequency ω_2 whereas previously it was at ω_1. The amplitudes are m_{A1} and m_{A2} and we take them to be approximately equal. Consider the following mechanism. Bearing in mind that the external oscillatory solution is the superposition of incident and transmitted waveforms, then consider the incoming waveform having an overlap of the original ω_1 waveform with a new ω_2 waveform. Once the transition region reaches the source particle, a similar overlap is propagated away from the particle in the transmitted waveform.

The incoming ω_2 waveform mixes with the incoming ω_1 waveform and then the transmitted outgoing ω_2 waveform mixes with the outgoing ω_1 waveform to produce a quasi-static 'steady state' waveform at $\Delta\omega$. It is easier to examine this at the time when the overlap reaches near to the centre of the source particle. This can be analysed as follows,

3.4 Photons

$$m_{BE} = \frac{2r_0}{r} A_m [|m_{A1}| \cos \omega_1 t + |m_{A2}| \cos \omega_2 t]^2$$
$$= \frac{2r_0}{r} A_m [|m_{A1}|^2 \cos^2 \omega_1 t + |m_{A2}|^2 \cos^2 \omega_2 t + |m_{A1}||m_{A2}|\{\cos(\omega_1 + \omega_2)t + \cos(\omega_1 - \omega_2)t\}]$$

where r_0 is some constant radius. The steady state components of the first two terms in the square bracket sum to provide the original potential variation,

$$\frac{r_0 A_m}{r}(|m_{A1}|^2 + |m_{A2}|^2) \cong \frac{2r_0 A_m |m_{A1}|^2}{r} = V$$

The $\omega_1 + \omega_2$ term is neglected in line with the low frequency filtering above and so m_{BE} now has an oscillatory component at the difference angular frequency $\Delta\omega = \omega_1 - \omega_2$. Since the steady state solutions are proportional to the square of the corresponding oscillatory solution at ω_1 or ω_2 and lead to $1/r$ potentials, as explained in the previous section, then the product of ω_1 and ω_2 oscillatory solutions leads to a $\Delta\omega$ solution with quasi-steady state (i.e. oscillatory at angular frequency $\Delta\omega$) fields with a $1/r$ dependence. The waveform at $\Delta\omega$ is determined by the length of the overlap region. A packet of carrier – Hermite function waveforms, i.e. the solution introduced in (3.1) and (3.2), can describe this. The waveform at $\Delta\omega$ formed by the mixing of ω_1 and ω_2 gravitational mass density waveforms is identified as the photon waveform. It has a v_0 parameter which is distinct from that of the omega waveform.

The connection between the simplified scheme presented here and the detailed solution of the Field Equations in the Book is dealt with in Appendix B. Appendix B shows that we can neglect the dispersion for the photon in the background with the result that the photon waveform travels in the background at velocity c. A further consequence is that the photon electric field E and magnetic field H only appear after entry to target particles. We have just mentioned two concepts, the electric field and the magnetic field, which we have not yet introduced. A digression is in order. Just as $V = m_{BE}$ external to a source particle, then we can introduce the magnetic vector potential

Photons

$$A = \mathbf{p}_{AE}/c^2$$

where $\mathbf{p}_{AE}$ is the external gravitational momentum density. This connection arises because it leads to the correct differential equation linking **A** to the current density. This means that we can introduce A, the magnitude of the magnetic vector potential within the photon waveform as follows.

The quasi-steady state (i.e. oscillatory) amplitude of the photon waveform due to the mixing process is

$$\frac{2r_0 A_m |m_{A1}||m_{A2}|}{r} \cong \frac{2r_0 A_m |m_{A1}|^2}{r} = V$$

This is the sum of the effect of incoming and outgoing waveforms. Associated with this amplitude is a gravitational momentum density cV - the factor c because of the steady state origin of these components. This must be transverse for reasons discussed in Chapter 3 in the Book. As we track the photon waveform out to a more remote radius, the travelling waveform component has an amplitude of $V/2$. The waveform is a quasi-steady state waveform, i.e. it is formed in the same way as the steady state potentials, but it is oscillatory. It induces a transverse oscillatory gravitational momentum density in target particles of $cV/2$. Thus the photon transverse magnetic potential amplitude is $A = V/2c$. In Cartesian co-ordinates we can write the magnetic vector potential as the photon waveform $A = B \exp(i\Delta\omega t - i\Delta k z)$ where Δk is the wave vector magnitude and B is a constant. The electric field and magnetic field are determined by the electromagnetic relationships,

$$E = -\frac{\partial A}{\partial t}$$

$$\mu_0 H = \frac{\partial A}{\partial z}$$

where the electric field is along x and the magnetic field along y as shown in Figure 3.6(a). You might ask why it is permissible to quote results from conventional electromagnetism. The formal answer is that the homogeneous and inhomogeneous Maxwell equations follow from the Field Equations together with the mathematics of electromagnetism.

3.4 Photons

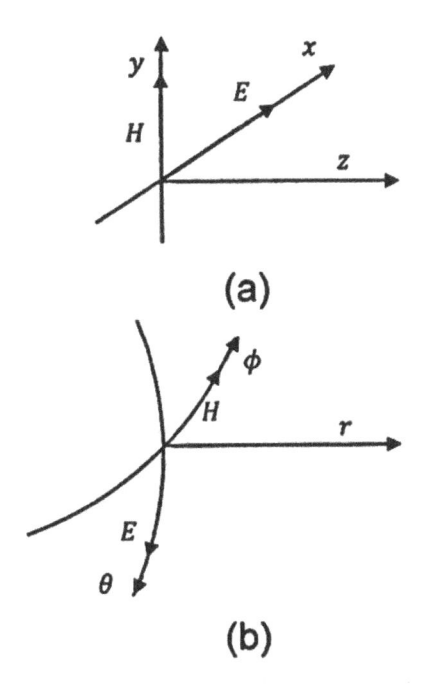

Figure 3.6 Photon waveforms (a) Cartesian co-ordinates. The electric field is along x, the magnetic field along y and propagation along z (b) Arrangement appropriate for dipole radiation using spherical co-ordinates in the vicinity of $\theta = 90°$. The electric field is along θ, the magnetic field along ϕ and propagation along r

It follows from the Field Equations in Chapter 1 (see Book) that

$$\frac{\partial^2 E}{\partial x^2} + \frac{\partial^2 E}{\partial y^2} + \frac{\partial^2 E}{\partial z^2} - \frac{1}{c^2}\frac{\partial^2 E}{\partial t^2} = -\frac{G}{\varepsilon_0}$$

$$\frac{\partial^2 G}{\partial x^2} + \frac{\partial^2 G}{\partial y^2} + \frac{\partial^2 G}{\partial z^2} - \frac{1}{c^2}\frac{\partial^2 G}{\partial t^2} = \frac{E}{\varepsilon_0}$$

where G is the gravitational field. Hence E and G are both travelling waves in the form of a summation of a product of a Hermite function and a sinusoidal carrier with n centred on n_0 as discussed in Section 3.3 and with their own v_0 as suggested above. Parallel equations apply to H and K. K is the gravnetic field which is the gravitational analogue of the magnetic field. By choosing a Type A waveform for E and H we remove G and K, and this leaves a travelling waveform with orthogonal E and H. This accounts for electromagnetic radiation with transverse electric and magnetic fields. Strictly we have described a solution which applies within particles; however this description is sufficient to discuss the transport of energy and gravitational mass. The propagation in the background between particles is discussed in Appendix B. The photon waveform along with a waveform we call the $\Delta\omega\, m$ waveform form the $\Delta\omega$ waveform (discussed in detail in Appendix B) and the latter is accompanied by a steady state gravitational mass density waveform whose integral is the gravitational mass transported along with the photon waveform. The quantum number n_0 is an integer which can change in increments of 1 within the Type A series of solutions keeping an E component at n_0. In the limit when $n_0 \to \infty$ the product sinusoid approximation tends to the infinitely long sinusoid solution of the conventional wave equation used in electromagnetism.

3.5 Photon transport of radiation energy and gravitational mass

In this section we deal with the 'energy' transported by a photon waveform. Energy is defined in the next chapter and formally we cannot use the concept before we introduce it. Nevertheless we can integrate the Poynting vector, whose magnitude is determined by the product of E and H, and draw conclusions. Specifically we can make a connection between the integral of the Poynting vector over the waveform and the electric potential. Whereas the electric potential is associated with entity P introduced in Section 3.2, the 'energy' involves a further entity Q, and as a result the fine structure constant involves P^2/Q as we shall see. We consider the integral of the Poynting vector which equals

$$\iint EH dS dt$$

where the integral is over the surface area S and over the time it takes for the waveform to pass through the surface. The details are in the Book and the case considered is for dipole radiation. The fields are

3.5 Photon transport of radiation energy and gravitational mass

depicted in Figure 3.6(b). The integral is over the surface of the sphere at radius r and over the time it takes for the waveform to pass at radius r. For the time being we need a name for this entity, and we shall call it the radiation 'energy'. It is shown in the Appendix (Section 3A.11) of the Book that the radiation 'energy' is given by

$$n_0 \hbar' \omega$$

where

$$\hbar' = \left(\frac{q^2}{4\pi\epsilon_0}\right)\left(\frac{Q}{2.09P^2}\right)$$

where q is the charge of the source particle and

$$Q = \sum_\nu \operatorname{sech}^4\left(\frac{\nu}{2\nu_0}\right)$$

and Q is plotted on Figure 3.4. Because the analysis is based on $n_0 >> 1$, then it is not possible to comment on whether n_0 should be replaced by $n_0 + 1/2$ and therefore whether the theory predicts the existence of zero point energy. The Fine Structure Constant is given by

$$FSC = \frac{1}{4\pi\epsilon_0}\left(\frac{q^2}{\hbar' c}\right) = \frac{2.09P^2}{Q}$$

and this is plotted on Figure 3.7.

Photons

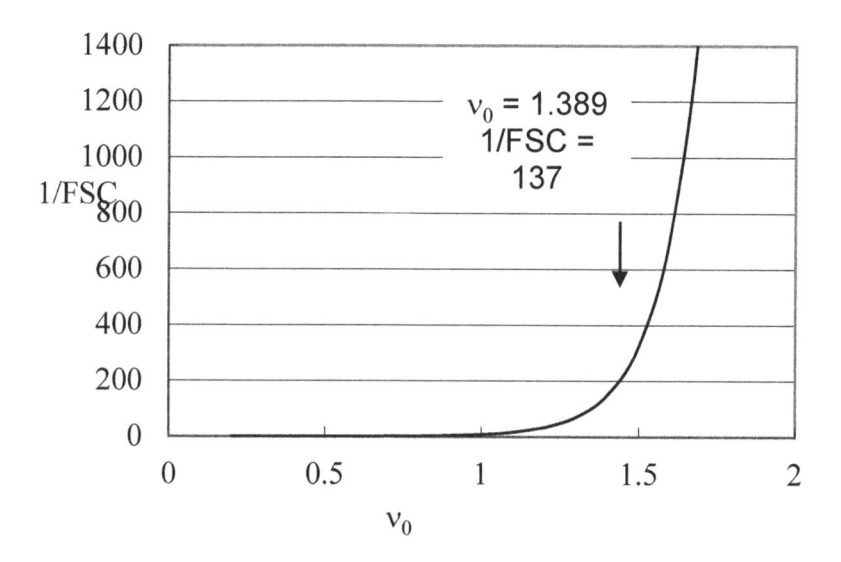

[Graph showing 1/FSC versus v_0, with annotations $v_0 = 1.389$, $1/FSC = 137$, x-axis labeled v_0 from 0 to 2, y-axis labeled 1/FSC from 0 to 1400]

Figure 3.7 1/FSC versus v_0

The Fine Structure Constant is usually denoted by α. However in the Book we have used $\alpha = 1/\epsilon_0$ and hence our use of the cumbersome initials FSC. Putting $h' = 2\pi\hbar'$, you may say that h' is Planck's constant. Well, we don't prove that to be the case until the next chapter.

There is a conclusion of major importance buried in these results. Since P^2 and Q are functions of v_0, then v_0 is determined if q^2 is fixed. As a consequence photons contain a distribution A_v versus v specific to a particular value of v_0 and therefore $|q|$. This means that charged particles are constrained to a particular value of $|q|$, since particles with other |charge| values cannot interact with existing photons. This applies to source and target particles. With $FSC = 1/137$ then $v_0 = 1.389$.

We mention above that the photon E and H waveforms are accompanied by a $\Delta\omega\,m$ waveform and this in turn is accompanied by a steady state gravitational mass density. We can calculate the total gravitational mass transported in a similar fashion to the calculation of the transported energy. The details are in the Book and it turns out that the increments in gravitational mass transported are also proportional to $\Delta\omega$. The photon travelling through the background enters the particle under consideration and an increment of the gravitational mass being transported is released.

3.6 Summary of photon properties

Similar to the omega waveforms based on Type A waveforms giving rise to photon waveforms by mixing, there are the Type B omega waveforms based on charge density giving rise to gravitational-gravnetic waveforms. Because they are based on charge density, the gravitational-gravnetic waveforms cannot carry gravitational mass. This is of major importance to the analysis of Chapter 5.

3.6 Summary of photon properties

The photon waveform delivers 'energy', accompanied by gravitational mass. The photon is not a point particle which somehow mysteriously gives rise to electric and magnetic potentials, and to oscillatory E and H fields. E and H are very real quantities when, for example, you probe microwave fields in a slotted line, and since in our model the photon has internal E and H fields, we can use our solutions to describe situations encountered in measurement. Remember that we have yet to introduce the idea of force, and we cannot yet interpret the electric field as the force per unit charge. Neither have we formally introduced the concept of energy.

We extract from the previous sections the following properties which are used in the next chapter:
(1) The electromagnetic radiation field is composed of a collection of photon waveforms, which are discrete waveforms each characterised by its own quantum number n_0 and angular frequency $\Delta\omega$.
(2) The photon waveform propagates conserving the value of $\iint EH dSdt$ and this equals $n_0 \hbar' \Delta\omega$.
(3) The outbound photon waveform can interact with another particle to produce a change in angular frequency $|\Delta\omega|$ where $|\Delta\omega|$ is the same for the new particle as for the previous one from which the photon waveform originated
(4) The |charge| is the same for all charged particles interacting with a common set of photon waveforms
(5) Photon waveforms are accompanied by gravitational mass, but they do not carry charge. The increment in gravitational mass is proportional to $\Delta\omega$ and to the increment in quantum number.

The various processes are summarised in Figure 3.8 and there is further discussion in Appendix B, Section B.6.

3.7 Summary

The problem tackled is the solution of the Field Equations to obtain the gravitational mass density and electric charge density fields external to a source particle. Solutions of the oscillatory Field Equations have been found in the form of travelling waveforms consisting of a superposition of products of a modified Hermite function wave packet and a sinusoidal carrier.

These solutions may be superimposed to provide the external oscillatory waveforms for a particle, the omega waveforms, and these give rise to the particle's steady state potentials. In order to obtain the required variation of the potentials with distance, a model is introduced in which the omega waveforms interact with particles in a background filling the space between individual particles.

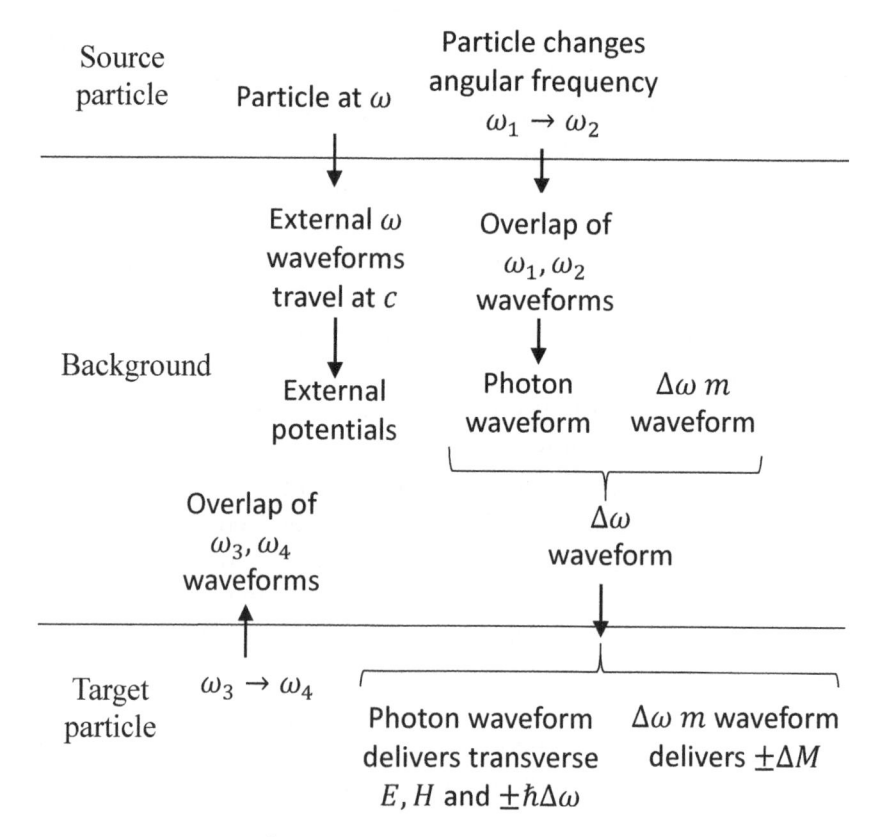

Figure 3.8 Summary of processes involving omega, photon, $\Delta\omega\, m$ and $\Delta\omega$ waveforms

3.7 Summary

When a particle changes its state from angular frequency ω_1 to ω_2 there is mixing between the ω_1 and ω_2 oscillatory waveforms resulting in a photon waveform at the difference frequency. The electric field is accompanied by a magnetic field and in the large quantum number limit, this corresponds to classical electromagnetic radiation. Similarly the gravitational field is accompanied by a gravnetic field and for large quantum numbers this can be described as gravitational-gravnetic radiation. A consequence of the connection between the external electric potential due to a source particle and the amplitude of the emitted photon waveform is that charged particles are constrained to a particular charge magnitude in order to interact with existing photons. It is shown that a photon waveform is accompanied by the transport of gravitational mass.

Chapter 4
Classical and quantum mechanics

4.1 Introduction

Classical mechanics has its origins in Newton's three laws of motion which state that (there are many versions):
1. Every body remains in its state of rest or uniform motion in a straight line, unless compelled to change that state by forces impressed upon it
2. The rate of change of momentum is proportional to the force impressed, and takes place in the direction in which that force acts
3. An action is always opposed by an equal reaction, that is, the forces that two bodies exert on each other are equal and opposite.

An example of a force was provided by Newton himself with his inverse square law for the gravitational force. Further examples are with Coulomb's inverse square law which describes the force by one electric charge on another electric charge and Ampere's law which describes the force one current exerts on another current.

There were two major extensions to classical mechanics in the twentieth century. The first is the introduction of relativity by Einstein, in two forms, special relativity and general relativity. This chapter presents largely non-relativistic results and the special relativity aspects of the new theory are dealt with in the Book. The connection of the new theory with general relativity, which concerns changes to conventional gravitational theory, is dealt with in Chapter 5. In the early part of the twentieth century quantum mechanics was developed to account for the properties of matter on the scale of individual atoms and the properties of electromagnetic radiation. In so doing it was clear that there is a close connection between quantum mechanics and classical mechanics.

4.2 The distributed particle

The origin for classical mechanics and quantum mechanics is proposed in this chapter. The problem tackled is to examine the effect that potentials induced by a source particle have on target particles and this requires that the solution scheme is extended to take account of the potentials. This chapter deals with charged particles. Models for neutral particles are introduced in Chapters 6 and 7. It will be seen that the requirement that the charge and gravitational mass induced within the particle by the potentials is provided out of the charge and gravitational mass already in the particle plays a crucial role in this investigation

4.2 The distributed particle

We are now going to investigate how particles can move when they encounter a potential emanating from another particle. Of course if an observer is moving with respect to a particle, the observer will say that the particle is moving. But what happens to that moving particle when it enters the potential field due to another particle? We know experimentally that the particle can be retarded or accelerated. So how do these effects arise in the new theory? Our first step is to see how we can describe a moving particle in the new theory.

Let's consider a one-dimensional situation where our particle has an oscillatory waveform $\rho_A(z)\exp(i\omega t)$. We can construct a new solution of the Field Equations by 'convolving' $\rho_A(z)$ with a wave function which we denote by ψ_D. We consider the case where $\psi_D = \exp(-ikz)$ and the new solution is in the form

$$[\exp(-ikz)] * [\rho_A(z)]\exp(i\omega t)$$

and similarly for the oscillatory gravitational mass density. We use z as the direction of motion so that the particle spin axis is aligned with this direction. The symbol $*$ denotes the convolution. If you do not know what a convolution is, no matter. It turns out that this expression can be a solution of the Field Equations so long as the particle angular frequency changes to a new value ω given by

$$\omega^2 - c^2 k^2 = \omega_0^2$$

(4.1)

Classical and quantum mechanics

So there can be a new solution provided that the angular frequency changes from ω_0 to ω, and the wave vector k is given by (4.1). The next step is to note that $\exp(i\omega t - ikz)$ is a travelling wave and we can superimpose a number of travelling waves each at a slightly differing ω and k to form a wave packet; an example is sketched in Figure 4.1.

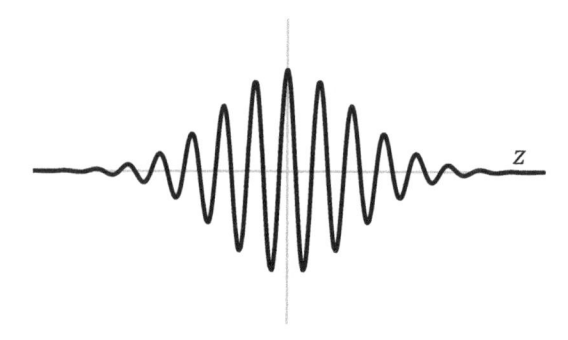

Figure 4.1 Example of a wave packet

This shows the situation versus distance at a particular instant of time. A similar picture is obtained when z is fixed and time varies as the wave packet passes the chosen point. Equation (4.1) is a dispersion formula which allows us to calculate a group velocity given by $w = d\omega/dk$ and so the wave packet moves along the z axis with a speed given by

$$w = c^2 k/\omega$$

which we can approximate by

$$w = c^2 k/\omega_0 \quad (4.2)$$

for speeds much less than c. We will not concern ourselves with the detail of the spread of the angular frequency but just say that ω_0 changes to a new value ω and we use (4.1) to find the required k. So if for some reason the angular frequency of a particle changes, it can result in particle motion. So a moving particle has changed its angular frequency from its rest value ω_0 to a new value ω given by (after manipulation of (4.1))

$$\omega - \omega_0 = c^2 k^2 / 2\omega_0$$

4.3 The introduction of Planck's constant

which we can express as

$$\omega - \omega_0 = \omega_0 w^2/2c^2 \qquad (4.3)$$

We shall return to this result shortly. $\exp(-ikz)$ distributes (i.e. spreads) the basic particle over an infinite distance, but the wave packet localises the distributed particle. Thus the wave function distributes the oscillatory solution over a volume larger than the original particle volume. The steady state densities are given by

$$(\psi_D \psi_D^*) * \rho_B \quad \text{and} \quad (\psi_D \psi_D^*) * m_B$$

and similarly $\psi_D \psi_D^*$ distributes the steady state gravitational mass and charge over a larger volume than the original particle volume. The integral of $\psi_D \psi_D^*$ over all space is unity ensuring that we retain the gravitational mass and electric charge of the original particle. We call the original particle the basic particle and after distribution we call the particle the distributed particle.

4.3 The introduction of Planck's constant

Next we introduce the external gravitational potential U and the external electric potential V due to a source particle. In Chapter 2, Sections 2.6 and 2.7, we point out that the gravitational mass commandeered by the electric potential inside a particle can only come from within the particle. This is a key point and is fundamental to why classical and quantum mechanics are the way they are as we shall see. First we introduce the average value m_{BM0} where

$$m_{BM0} \mathcal{V} = M_0$$

and $\mathcal{V}$ is the particle volume. As a consequence we can write using equation (2.10),

$$m_{BM} + V = m_{BM0}$$

where m_{Bm} is the new average steady state gravitational mass density. Similarly the charge commandeered within a particle by the gravitational potential can only come from the charge within the particle and so with $\rho_{BM0} \mathcal{V} = q$,

$$\rho_{BM} + U = \rho_{BM0}$$

Classical and quantum mechanics

We consider the case where U is small compared to ρ_{BM} and V is small compared to m_{BM}. We equate the product of the two left hand sides to the product of the two right hand sides and then multiply by the particle volume $\mathcal{V}$,

$$(\rho_{BM} + U)(m_{BM} + V)\mathcal{V} = \rho_{BM0} m_{BM0} \mathcal{V}$$

and neglecting the UV term we have

$$(\rho_{BM} m_{BM} - \rho_{BM0} m_{BM0})\mathcal{V} + qV + M_0 U = 0$$

We are now seeing the familiar expressions for potential energy appearing – but that is racing ahead. We manipulate this equation by introducing m_{AM} and ρ_{AM} where

$$m_{BM} = A_m m_{AM} m_{AM}^*$$

$$\rho_{BM} = A_\rho \rho_{AM} \rho_{AM}^*$$

and similarly the original values m_{AM0} and ρ_{AM0} are related to m_{BM0} and ρ_{BM0}. It turns out that the product $\rho_{AM} m_{AM}^*$ depends on the angular frequency, and so the first term on the left hand side is proportional to the change in angular frequency $\Delta\omega = \omega - \omega_0$. The details are in Appendix 4A of the Book. So we write the equation as

$$\hbar \Delta\omega + qV + M_0 U = 0$$

(4.4)

where $\hbar$ is a constant and it is given by (see the Book Chapter 4, Appendix 4A, for the derivation)

$$\hbar = -\frac{M_0 A_m \rho_{AM0} m_{AM0}^*}{\omega_0}$$

(4.5)

Note that the expression contains a minus sign. This is because ρ_{AM0} and m_{AM0} are complex quantities and are in anti-phase, and the expression delivers a positive value for $\hbar$. Putting $h = 2\pi\hbar$ then h is Planck's constant. $\hbar$ is to be treated as a fundamental constant and the set of constants in Chapter 2 becomes c, ε_0, A_m, A_ρ and $\hbar$.

We can now make connection with the results in Chapter 3. This is a summary of a more detailed analysis in the Book. When a particle changes its angular frequency due to the absorption of a photon, the photon brings in 'energy' $\hbar(\omega - \omega_0)$ and we can see that we require

4.4 The classical equations of energy and motion

$\hbar'$, introduced in Chapter 3, to equal $\hbar$. The magnitude of electric charge is the same for all charged particles. The gravitational mass transported by photons is proportional to the frequency of the photon and so the change in gravitational mass of the source particle (or target particle) is proportional to the change in angular frequency of the source particle (or target particle) i.e. $M - M_0$ is proportional to $\omega - \omega_0$. So we can write $M_0 = A\omega_0$ or more generally $M = A\omega$ where we call A the particle gravitational mass constant. We can express $\hbar$ in the form

$$\hbar = -AA_m K \tag{4.6}$$

where $K = \rho_{AM0} m^*_{AM0}$ is a constant. These simple particle properties are used throughout the Book and in this book to determine various particle parameters in terms of the fundamental constants.

4.4 The classical equations of energy and motion

Define inertial mass by

$$\mathcal{M}_0 c^2 = \hbar \omega_0 \tag{4.7}$$

that is we have chosen to call $\hbar\omega_0/c^2$ the inertial mass. Note that since $M_0 = A\omega_0$ then the gravitational mass measured in Coulombs is proportional to the inertial mass measured in kilograms. This has enormous consequences which become apparent as the development proceeds. All three quantities, c, $\hbar$ and ω_0 already exist as physical entities within the new theory and hence the inertial mass is introduced into the new theory as a well defined physical entity. When w is much less than c then using (4.3) and (4.7), (4.4) becomes

$$\frac{1}{2}\mathcal{M}_0 w^2 + qV + M_0 U = W \tag{4.8}$$

We have introduced W on the right hand side which is a generalisation of (4.3) discussed in the Book. We call the first term the kinetic energy and we call the terms qV and $M_0 U$ the electric and gravitational potential energies respectively. We can now say that the external potentials V and U are the potential energies per unit charge and

Classical and quantum mechanics

gravitational mass respectively. So as particles gain increased potential energies, the kinetic energy must reduce and the particle retards. The greater the rate of change of the potential with distance then the greater the rate of change with time of the speed. We can write this in the form of the equation of motion,

$$F \equiv \frac{d}{dt}(\mathcal{M}_0 w) = -\frac{\partial}{\partial z}(qV + M_0 U) = qE + M_0 G$$

(4.9)

where we define the particle inertial momentum as $\mathcal{M}_0 w$ and F is the force, defined as the rate of change of inertial momentum. The electric field E and the gravitational field G are given by

$$E = -\frac{\partial V}{\partial z}$$

$$G = -\frac{\partial U}{\partial z}$$

We can identify E as the force per unit charge and G as the force per unit gravitational mass. Note that the force is defined acting on the whole particle and is expressed in terms of external potentials from other particles. Appropriate integration of (4.9) leads back to the equation of energy. In so doing we multiply the forces on the right hand side by elements of distance and then add them up. This is obtaining the work done, and is strictly our definition of energy. So the terms in (4.8) are themselves various forms of energy as we have already discussed.

When the right hand side of (4.9) is zero, and the inertial momentum is not changing, this is just as required by Newton's first law. If we describe the effect of the changes with distance of the potential energies as the application of an electric force and a gravitational force, then equation (4.9) says that the rate of change of inertial momentum is equal to the applied force, which is what is required by Newton's second law. If we have a pair of particles, and there is a no external force acting on the system, since there is no change in the system total inertial momentum then the forces of each particle on the other are equal and opposite as required by Newton's third law.

We refer to $\mathcal{M}_0$ and $\mathcal{M}_0 c^2$ as the rest inertial mass and the rest energy. In classical mechanics the inertial mass remains at $\mathcal{M}_0$ but in

4.4 The classical equations of energy and motion

special relativity the inertial mass increases from $\mathcal{M}_0$ to $\mathcal{M}$ as the velocity increases. The derivation of special relativity mechanics is dealt with in the Book.

The potential at a distance r from a point charge q_1 is given by $q_1/4\pi\varepsilon_0 r$ and this leads to the force on a charge q_2, i.e. Coulomb's law,

$$F = \frac{q_1 q_2}{4\pi\epsilon_0 r^2}$$

The Book deals with magnetism and this leads to Ampere's law, which describes the force between current elements. The formal connection with electromagnetism is completed in the Book. It is shown that there is a force arising from a component of the electric field proportional to the rate of change of the magnetic vector potential – this leads to Faraday's law of induction. The new theory, via Maxwell's equations, introduces the Maxwell displacement current. The gravitational potential due to a point gravitational mass M_1 is $-M_1/4\pi\varepsilon_0 r$ and so the force on a gravitational mass M_2 due to M_1 is

$$F = -\frac{M_1 M_2}{4\pi\epsilon_0 r^2}$$

Since the inertial masses are proportional to the gravitational masses, we can recast this result as

$$F = -\frac{\mathcal{G} M_1 M_2}{r^2}$$

where

$$\mathcal{G} = \frac{1}{4\pi\epsilon_0}\left(\frac{A^2 c^4}{\hbar^2}\right)$$

(4.10)

and so we have obtained Newton's law of gravitation. The expression for the gravitational constant is used in Chapter 8 to predict a theoretical value for comparison with the measured value. The Book also deals with the gravitational analogue of the magnetic field, the gravnetic field, and the new theory predicts a gravitational version of Ampere's law describing a force between moving mass elements.

Classical mechanics via Newton's laws leads to the design techniques underpinning civil engineering, for example with bridge

Classical and quantum mechanics

and tower block design, and mechanical engineering, for example with the design of aircraft structures, transmission systems, car safety and wind turbines. The concept of energy and its manifestation in various forms underpins the subject of thermodynamics and its application to internal combustion engines, refrigerators, heat pumps and chemical reactions. Newton's law of gravitation is fundamental to the calculation of satellite orbits and the understanding of the trajectories of shells and ballistic missiles. The extension of the theory developed above to include the magnetic field leads to the electromagnetic theory which underpins the design of electrical circuits, electronics and antennae and hence the design of communications systems, radar systems and electrical power generation and distribution.

4.5 Schrödinger's equation

Using expression (4.2) the equation of energy (4.8) becomes

$$\frac{\hbar^2 k^2}{2M_0} + M_0 U + qV = W$$

Putting

$$\psi_D = \cos kz$$

$$\frac{d^2\psi_D}{dz^2} = -k^2\psi_D$$

and so

$$-\frac{\hbar^2}{2M_0}\frac{d^2\psi_D}{dz^2} + (M_0 U + qV)\psi_D = W\psi_D \tag{4.11}$$

This is the one dimensional form of Schrodinger's equation and it is at the heart of introductory accounts of quantum mechanics. There is also angular momentum associated with the particle, which we call spin. It is shown in Appendix 4C in the Book that the spin parameter s introduced in Chapter 2 is related to the particle angular momentum in the form $s\hbar$ in the same way that the linear inertial momentum of a particle can be expressed as $\hbar k$. Equation (4.11) can be generalised to a three-dimensional form and this can be solved for the case of an electron held in a potential well in the vicinity of an atomic nucleus to

4.6 Summary

obtain the allowable values of W which correspond to the energy levels within atoms of atomic electrons. As a consequence quantum mechanics underpins our understanding of atomic theory, solid state physics, chemistry via the understanding of the nature of chemical bonding between atoms, and the laws of thermodynamics via statistical mechanics. These in turn underpin the design of electronic devices and microelectronics, microwave devices and optical devices of which the laser is a major example.

A fundamental point regarding quantum mechanics in the new theory needs to be reiterated and underlined. In conventional quantum mechanics the particle is essentially point-like, and the wave function in the form $\psi_D \psi_D^*$ is interpreted as the probability that the particle is at the selected point. In the new theory $\psi_D \psi_D^*$ smears the steady state particle charge and gravitational mass. The particle is no longer the original basic particle – it is physically occupying a larger volume than before. This is not to say that probability does not play a role in quantum mechanics. How probability arises in quantum mechanics, according to the new theory, is dealt with in Chapter 12.

4.6 Summary

The original or basic particle description can be modified by a wave function which distributes the particle's electric charge and gravitational mass over a larger volume. There can be a new solution of the Field Equations so long as the particle changes its angular frequency. The new solution can also be cast in the form of a travelling wave packet. When the particle is immersed in the potential fields from other particles, the particle frequency again needs to change and the analysis leads to an expression for Planck's constant. Combining these two features results in an equation of energy, an equation of motion and Newton's laws of motion. These lead to Coulomb's law and Newton's law of gravity. This establishes the basis for classical mechanics, and also provides an expression for the gravitational constant. It is shown that the wave function is a solution of Schrödinger's equation and when this is generalised to a three dimensional form, it provides the starting point for quantum mechanics.

Chapter 5

Gravitation

5.1 Introduction

As should be very evident, this theory is not based upon Einstein's theory of general relativity. However general relativity theory makes many predictions which accord with observation and experiment. It is the purpose of this chapter to demonstrate that the new theory makes the same predictions as general relativity, at least as far as those to do with the solar system are concerned.

It has already been shown in the previous chapters that the new theory leads to an account of classical mechanics, electromagnetism and quantum mechanics, and to the existence and some properties of fundamental particles and photons. The purpose of the present chapter is to extend the theory to show that some of the major predictions of general relativity which are consistent with experiment and observation also follow from the new theory. These predictions concern the advance of the perihelion of planets, the 'red shift' observed with clocks on and above the Earth's surface and the bending of radiation in a gravitational field. These predictions are obtained from the new theory independently of the concepts and assumptions of general relativity.

5.2 Motion in a gravitational field

We have seen that a charged particle accelerated by an electric field absorbs energy from a photon or photons, and that gravitational mass is delivered to the particle. However a particle accelerated in a gravitational field cannot interact with a gravitational-gravnetic (i.e. Type B) waveform in the same way because the waveform does not carry gravitational mass. As a consequence it is concluded that the accelerating particle needs to maintain its angular frequency constant. This is the principle that the new theory uses in deriving the results obtained in general relativity without making the assumptions underlying relativity. This is of major significance for the status of the

5.3 The effect of the gravitational potential on orbital motion

new theory, in distinction from other contenders for explaining the fundamentals of physics which assume general relativity as a fundamental constituent. To ensure that this principle does not escape the reader's attention we repeat it.

> A particle, moving in a gravitational field only, maintains its angular frequency constant.

This principle is first applied to the motion of a massive body in orbit around the sun leading to the prediction of the advance of the perihelion. The adjustment of the angular frequency as a function of the gravitational potential leads to the prediction of the gravitational red shift. Attention is then directed at the effect of a gravitational potential on the behaviour of electromagnetic radiation.

5.3 The effect of the gravitational potential on orbital motion

The following analysis of planetary motion is based on two steps. The first concerns the principle that an orbiting body's particles maintain their angular frequencies constant. The second step exploits this principle to obtain a modified equation of energy for an orbit from which the advance of the perihelia of the planets can be predicted.

We start with the body in free space. Split the body into positively and negatively charged constituents, say electrons and nuclei, and transfer each constituent into orbit and then reassemble the body. We need charged constituents in order to manipulate them notionally with non-gravitational fields. The analysis refers to one of the particles in these constituents. It follows the particle on a radial path towards the sun. The particle is brought to rest and then accelerated into orbit. All the behaviour is as observed by an observer in free space and stationary with respect to the sun.

Let the particle in free space have an angular frequency ω_0 and inertial mass M_0. Allow the particle to enter the sun's gravitational field, and it falls along a radius towards the sun. It might be expected that when a particle is accelerated in a gravitational field there is an interaction with a transverse gravitational-gravnetic wave with a Type B waveform analogous to the interaction with a Type A waveform, i.e. a photon, in an electric field. However because Type B waveforms cannot transport gravitational mass, the gravitational mass of the particle will remain at the free space value M_0. Since $M_0 = A\omega_0$ then

Gravitation

the angular frequency will remain at ω_0. However the equation of energy requires that one quantum of energy is deposited in the particle so that

$$\hbar\Delta\omega + M_0 U = 0 \tag{5.1}$$

where $\Delta\omega$ is the positive change in angular frequency. However since there can be no transport of gravitational mass into the particle and therefore no change in particle angular frequency, there can be no signalling into the surrounding space of such a change. Hence there must be a quantum increase in the outbound radiation to compensate for the quantum of energy delivered. This can come from the particle internally so that its rest inertial mass is reduced by

$$\frac{\Delta\omega \mathcal{M}_0}{\omega_0} = -\frac{M_0 U}{c^2}$$

and the rest inertial mass becomes, where $\mathcal{M}_S$ is the inertial mass of the sun and r is the distance from the sun,

$$\mathcal{M}_0 \left(1 + \frac{M_0 U}{\mathcal{M}_0 c^2}\right) = \mathcal{M}_0 \left(1 - \frac{G\mathcal{M}_S}{rc^2}\right)$$

and the corresponding change in the particle rest angular frequency is $\gamma\omega_0$ where

$$\gamma = -\frac{G\mathcal{M}_S}{rc^2} \tag{5.2}$$

showing that γ depends only on the gravitational potential energy, i.e. $\gamma = M_0 U / \mathcal{M}_0 c^2$.

When the angular frequency is ω_0 the inertial mass is constant at $\mathcal{M}_0$. The inertial momentum is defined as $\mathcal{M}_0 w_r$ where w_r is the radial velocity, and the force applied is defined by $(d/dt)(\mathcal{M}_0 w_r)$. The work done and therefore the kinetic energy is $\mathcal{M}_0 w_r^2 / 2$. Thus the equation of energy (5.1) becomes

$$\frac{1}{2} \mathcal{M}_0 w_r^2 - \frac{G\mathcal{M}_S \mathcal{M}_0}{r} = 0 \tag{5.3}$$

5.3 The effect of the gravitational potential on orbital motion

Using (5.2) this equation can be converted into a frequency balance equation

$$\frac{1}{2}\left(\frac{w_r^2}{c^2}\right)\omega_0 - \frac{GM_S}{rc^2}\omega_0 = 0$$

(5.4)

where the first term is the angular frequency change associated with the kinetic energy.

Next we are going to insert the radial moving particle into an orbit around the sun. It involves the following steps and involves electric forces as well as gravitational forces. The first step is to bring the particle to rest at a distance r from the sun by a collision with another particle or body. In the second step, in disengaging from the collision, the particle is accelerated to a speed $r\, d\phi/dt$, where ϕ is the spherical co-ordinate angle. The particle is inserted into orbit. At the same time the energy constant is changed to W' derived from an internal electric potential (see Chapter 4). These steps are as follows.

By collision or collisions involving electric forces, bring the particle to rest. The rest mass is $M_0(1 - GM_S/rc^2)$ as explained above. Remove energy $|W'|$ and accelerate the particle into orbit by gravitational means. Put the new inertial rest mass in orbit in the form $M_0(1 - X)$ where X is to be determined. Since the angular frequency remains constant during orbit insertion, then the kinetic energy is in the form as in equation (5.3) and is given by

$$\frac{1}{2}M_0 r^2 \dot{\phi}^2 (1 - X)$$

The reduction in angular frequency below ω_0 due to the rest mass $M_0(1 - X)$ plus the increase in angular frequency due to the kinetic energy must equal the reduction in the angular frequency due to the rest mass $M_0(1 - GM_S/rc^2)$ and the change in angular frequency due to the energy change W'. Thus the frequency balance equation, analogous to (5.4), is

$$-X\omega_0 + \frac{r^2\dot{\phi}^2(1-X)}{2c^2}\omega_0 = -\frac{GM_S}{rc^2}\omega_0 + \frac{W'}{\hbar}$$

(5.5)

Gravitation

For the purposes of finding X approximate the kinetic energy by

$$\frac{1}{2}M_0 r^2 \dot{\phi}^2$$

and substitute for the kinetic energy contribution in (5.5) from the classical result for a circular orbit, obtained by introducing W' into (5.3)

$$\frac{1}{2}M_0 r^2 \dot{\phi}^2 - \frac{GM_S M_0}{r} = W'$$

and we have

$$X = \frac{2GM_S}{rc^2} \tag{5.6}$$

This is approximate because of the way it has been obtained. We show below that this result is exact. In orbit the particle is under gravitational control. The gravitational mass and the angular frequency are constant, and so the inertial mass is constant at some value $\mathcal{M}_W$. Substituting for X from (5.6) into equation (5.5), multiplying by $\mathcal{M}_W$ and with inclusion of a small amount of radial kinetic energy resulting from a gravitational nudge in the form $(1/2)\mathcal{M}_W \dot{r}^2$ and with

$$W = \frac{W' \mathcal{M}_W}{\mathcal{M}_0}$$

the equation of energy for the particle in orbit becomes

$$\frac{1}{2}\mathcal{M}_W \dot{r}^2 + \frac{1}{2}\mathcal{M}_W r^2 \dot{\phi}^2 \left(1 - \frac{2GM_S}{rc^2}\right) - \frac{GM_S \mathcal{M}_W}{r} = W \tag{5.7}$$

It might be said that we could have multiplied (5.5) by some other value of inertial mass. However by multiplying by $\mathcal{M}_W$ (5.7) is then in a form which reduces to the classical approximation when the change in rest mass with gravitational potential is removed and this allows alignment of the results here with the classical results obtained in Chapter 4. Also we notice that the expression for the gravitational potential energy in (5.7) is exactly what it should be for a constant inertial mass of $\mathcal{M}_W$ and so the expression for X above is exact. It is shown in the Book, Chapter 5, that

5.4 The gravitational red shift

$$r^2\dot{\phi} = J \tag{5.8}$$

where J is an orbital constant.

If the time in (5.7) and (5.8) is replaced by the proper time in general relativity theory, they are identical to the equations obtained in Einstein's general theory of relativity (Kenyon 1990 p90) and this connection is discussed in detail in Appendix 5B in the Book. However we have no need to introduce the concept of proper time and Equations (5.7) and (5.8) leads to the orbital equation, where $u = 1/r$,

$$\frac{d^2u}{d\phi^2} + u - \frac{GM_S}{J^2} = \frac{3GM_Su^2}{c^2} \tag{5.9}$$

The term in $1/J^2$ is derived from the potential energy term in (5.7). The term on the right hand side is zero for the classical Newtonian case. It is the presence of this term which leads to an expression for the advance of the perihelion in one revolution which is in close agreement with observation. The behaviour of Mercury is predicted to a high accuracy and agreement is obtained within or comparable to the measurement error for Venus and the Earth. It must be emphasised that we have obtained these results from the new theory independently of the concepts and assumptions of general relativity. Note also that the new theory points to there being detailed physics involved in the internal description of particles on insertion into orbit.

5.4 The gravitational red shift

Equation (5.2) is modified when on or above the Earth's surface to

$$\gamma = -\frac{GM_S}{rc^2} - \frac{GM_E}{Rc^2}$$

where R is the distance from the centre of the Earth and M_E is the Earth's inertial mass. This means that the particles in a source of radiation or clock in the vicinity of the Earth have their rest inertial masses reduced. So angular frequencies ω_0 and ω are shifted to lower angular frequencies $\omega_0(1 + \gamma)$ and $\omega(1 + \gamma)$ which are dependent on the height above the Earth's surface. This also applies to the difference frequency $\omega - \omega_0$ and therefore to emitted radiation. This result is also obtained in general relativity (Kenyon 1990 p17) and has been amply verified by experiment. Again it must be emphasised that this result has been obtained from the formalism independently of the concepts and assumptions of general relativity.

5.5 Bending of light in a gravitational field

The radiation with energy $\hbar\omega$ and gravitational mass $A\omega$ has the properties of a particle of inertial mass $M = \hbar\omega/c^2$ and satisfies the orbital equation (5.9) where $u = 1/r$. Since the radiation is travelling at around c, then the potential energy term is small compared to the kinetic energy terms and it can be neglected, resulting in

$$\frac{d^2u}{d\phi^2} + u = \frac{3\mathcal{G}\mathcal{M}_s u^2}{c^2}$$

It is a standard piece of analysis to derive the deflection of radiation just grazing the sun's surface given by

$$\Delta\phi = \frac{4\mathcal{G}\mathcal{M}_s}{bc^2}$$

where b is the distance of closest approach (see Kenyon 1990 pp 93-94). The predicted deflection of 1.75 arcsec is in agreement with observation for both light and radio waves. Again the point needs to be made that the new theory has obtained these results independently of the concepts and assumptions of general relativity.

5.6 The status of general relativity within the new theory

The new theory predicts experimental results predicted by general relativity as described in this chapter. However the analysis in general relativity is developed using a co-ordinate space that is curved with respect to the flat space of the new theory. Also general relativity is based on its own postulates. So can general relativity be derived from the new theory? This is investigated in Appendix 5B of the Book. Although the general relativity postulates are not derived in their general form, this appendix shows that intermediate results are derivable up to the complexity of a Schwarzschild metric.

5.7 Conclusions

The new theory leads to significant aspects of general relativity. Models for fundamental particles are developed in the following chapters. The new theory takes into account gravity in modelling fundamental particles but this is not the same as having to incorporate general relativity into particle models from the outset. Again we emphasise that these results (advance of perihelion, red shift and the bending of light) have been obtained without the extra assumptions of general relativity.

In 2011 it was reported by the OPERA consortium (Adam et al 2011) that muon neutrinos had been observed travelling faster than the speed of light. This claim was subsequently withdrawn (Adam et al 2012). The conclusion either way has implications for the new theory and this is discussed in Appendix C, with the following results. It is concluded that in the new theory the velocity of radiation in a local gravitational potential is measured to be c by local observers moving uniformly with respect to each other. This is inconsistent with what is measured by a distant observer in zero gravitational potential using radiation which passes through the local region. The distant observer concludes that the radiation is moving through the local region at a velocity less than c. This is at odds with the Second Postulate, Chapter 1, Section 1.2. This can be resolved by referring to local observers in the Second Postulate. This does not affect the results from Chapters 1 to 4 which involve only local observation. In fact the change ensures that Chapters 1 to 4 apply under the local circumstances introduced above. We have therefore deduced that special relativity applies locally within a frame in free fall in the local gravitational field. Indeed we have derived the strong equivalence principle - that the laws of physics apply in such frames in free fall - thus obtaining a major principle underlying general relativity. When we take into account the discussion in Appendix 5B in the Book which concludes that the results of general relativity up to the complexity of the Schwarzschild 2-D metric can be derived from the new theory, then we can claim that a form of general relativity follows from the new theory. The details of this form of general relativity need to be worked through for comparison with conventional general relativity. In Chapter 13 of the Book it is conjectured that there is no need for fundamental postulates or fundamental constants. These conclusions are not affected by special relativity applying locally, as discussed in Chapter 18 in this book.

5.8 Summary

It might be expected that when a particle is accelerated in a gravitational field there is an interaction with a Type B waveform analogous to the interaction with a Type A waveform in an electric field. However, because Type B waveforms cannot transport gravitational mass, the gravitational mass of the particle and its angular frequency will remain at their free space values. As a consequence it is concluded that there is a shift in a particle's angular frequency when immersed in a gravitational potential in order to compensate for the angular frequency change due to the particle's motion. It is shown that these results lead to a modified equation of energy for orbital motion and to the prediction of the advance of the perihelion of planets identical to that in Einstein's theory of general relativity. The predicted shift in a particle's angular frequency when immersed in a gravitational potential leads directly to the prediction of the red shift of oscillators in the vicinity of the Earth's surface. An orbital equation is obtained for electromagnetic radiation and this leads to the prediction of the observed bending of radiation in a gravitational field. These various predictions have been obtained from the new theory independently of the concepts and assumptions of general relativity. A case is made in Appendix C that a form of general relativity follows from the new theory.

Chapter 5 references

Kenyon I R 1990 *General Relativity* (Oxford)
OPERA collaboration Adam T et al, 22 Sep 2011 *Measurement of the neutrino velocity with the OPERA detector in the CNGS beam* arXiv: 1109.4897 hep-ex
OPERA collaboration Adam T et al, 12 July 2012 *Measurement of the neutrino velocity with the OPERA detector in the CNGS beam* arXiv: 1109.4897 hep-ex v4

Chapter 6

Baryons and mesons

6.1 Introduction

The standard model classifies matter particles as hadrons or leptons. We deal with hadrons in this chapter and with leptons in the next. Hadrons are either baryons or mesons. Mesons have spin 0 or 1 and baryons have spin 1/2 or 3/2. Quarks are a central features of the standard model and mesons contain two quarks and baryons contain three quarks. In later chapters we show how quarks arise in the new theory and how models of various particles can be constructed from them.

Chapter 2 introduces a particle model and therefore shows how the existence of fundamental particles arises in the new theory. An oscillatory solution scheme for the Field Equations is presented which leads to categorisation of particles based on their spin angular momentum. The general problem for determining particle solutions is reduced to one of finding solutions of the steady state scalar Field Equations, given the oscillatory solutions.

In this chapter and the two subsequent chapters we apply the general particle scheme to particular particles and show that theoretical expressions for various particle parameters can be obtained. In this chapter baryons and mesons are investigated. In Chapter 7 leptons are investigated and in Chapter 8 it is demonstrated how to obtain theoretical values for major constants and many particle parameters from the five fundamental constants. The analysis is far from rigorous. It involves approximations and appeals to qualitative arguments. The identification with actual particles is done on the basis of the angular frequencies which result from the analysis together with the spin quantum numbers in anticipation of these being put into correspondence with the experimental values in Chapter 8. The product from the three chapters is summarised in a set of structures proposed for fundamental particles depicted in Figure 8.1. In Chapter 9 and the following chapters the theory is extended to include the existence of quarks.

This chapter proposes a model for a spherical particle model

Baryons and mesons

resulting from solution of the oscillatory and steady state Field Equations; the solutions lead to structures based on a series of concentric spherical shells. This is adapted to the proton and neutron resulting in structures with multiple cylindrical shells. The proton model provides key steps in linking together various fundamental constants and these are used in Chapter 8 to predict values for the charge on the electron and the gravitational constant. The shell model is also applied to the kaon. A different solution of the Field Equations leads to a proposed structure for the pion. A scheme is proposed to account for anti-particle and neutral counterparts.

6.2 Particle parameters

In Chapter 3 it is shown that the charge of charged particles is $\pm q$ where q is a constant. In Chapter 4 it is shown that the gravitational mass is proportional to the particle angular frequency, i.e. $M_0 = A\omega_0$. It is also shown that that the product $\rho_{AM0} m^*_{AM0}$ is a constant K, where ρ_{AM0} and m_{AM0} are the values for the particle at rest. Strictly this last relationships applies to a subset of particles – further detail is in Appendix J. This chapter is concerned only with particles at rest and the subscript 0 will be dropped from ρ_{AM0} and m_{AM0}. Remembering that

$$q = A_\rho \rho_{AM} \rho^*_{AM} \mathcal{V} = \rho_{BM} \mathcal{V}$$

$$M_0 = A_m m_{AM} m^*_{AM} \mathcal{V} = m_{BM} \mathcal{V}$$

it is straightforward to show that the variation of parameters with angular frequency for particles in the subset is as follows:

$$|m_{AM}| = \left(\frac{A_\rho K^2 A}{A_m q}\right)^{1/4} \omega_0^{1/4}$$

(6.1)

$$|\rho_{AM}| = \left(\frac{A_m q K^2}{A_\rho A}\right)^{1/4} \omega_0^{-1/4}$$

(6.2)

$$\mathcal{V} = \left(\frac{Aq}{A_\rho A_m K^2}\right)^{1/2} \omega_0^{1/2}$$

(6.3)

6.3 The spherical particle shell model

$$\frac{|m_{AM}|}{|\rho_{AM}|} = \left(\frac{A_\rho A}{A_m q}\right)^{1/2} \omega_0^{1/2} \tag{6.4}$$

and we make use of these results in this chapter and subsequently.

6.3 The spherical particle shell model

We now turn attention to the construction of a solution of the steady state Field Equations given a solution of the oscillatory Field Equations. We consider the case of the spherical particle which arises from Group A when $s = 0$. Since there is no dependence of ρ_A and m_A on θ or ϕ there is a radial variation only and this corresponds to a spherical particle. The oscillatory solution can be put into the form (see Book, Section 6A.3),

$$\rho_A = \frac{\sqrt{2} B_{\rho A}}{(r\pi x)^{1/2}} \left[\sin x + i(1 + b_\rho') \cos x\right] \tag{6.5}$$

where the Bessel functions in the Book have been replaced by $\sin x$ and $\cos x$ and where $x = \omega_0 r/c$ and where $B_{\rho A}$ and b_ρ' are constants. x expresses a radius in units of c/ω and is used extensively here and in later chapters. It might be thought that there is a simple solution with $b_\rho' = 0$. The solution in this case is of the form $D \exp(-ix)/r$ where D is a constant. However there is a problem in that this contains a $1/r$ factor and consequently the steady state solution given by $\rho_{B1} = A_\rho \rho_A \rho_A^*$ (neglecting ρ_{B2}) is proportional to $1/r^2$. This does not satisfy Laplace's equation, see Section 2.5, whereas it will if it is proportional to $1/r$. This problem can be tackled by allowing small values of b_ρ' as follows. We have for the steady state charge density,

$$\rho_{B1} = A_\rho \rho_A \rho_A^* = \frac{A_\rho B_{\rho A}^2}{r} \left[\frac{2}{\pi x}(1 + b_\rho' + b_\rho' \cos 2x)\right]$$

where we have dropped terms in $b_\rho'^2$. The condition for the expression in the square brackets to be independent of x can be approximated by the condition that its gradient be zero at some point x_0 and there will be some range in x either side of x_0 over which the departure from a $1/r$ variation is acceptable. The $1/x$ factor results in a downward

Baryons and mesons

gradient as x increases. This can be counteracted by an upward gradient due to $b_\rho' \cos 2x$. This latter gradient is at a maximum when $\cos 2x = 0$, see Figure 6.1, and this occurs at intervals in x of $\pi/2$.

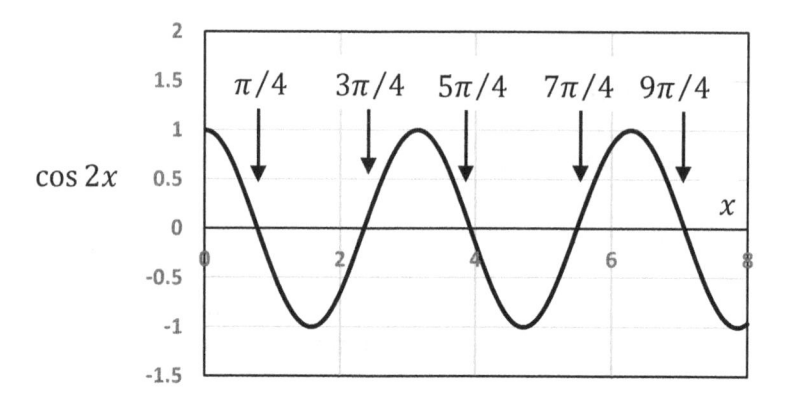

Figure 6.1 $\cos 2x$ versus x. The positive gradients are greatest for $x = 3\pi/4, 7\pi/4$ etc., the negative gradients are steepest at $x = \pi/4, 5\pi/4, 9\pi/4$ etc.

This condition provides the values of x_0 around which the greatest extents of the acceptable ranges in x occurs. When $x = n\pi - \pi/4$ where n is an integer, b_ρ' is required to be positive and

$$b_\rho' = \frac{1}{2x_0 - 1}$$

and when $x = n\pi - 3\pi/4$ we require b_ρ' to be negative and

$$b_\rho' = -\frac{1}{2x_0 + 1}$$

A similar treatment applies to the oscillatory and steady state gravitational mass densities but complicated by extra terms – the details are in the Book. The results from above are summarized in Table 6.1 and illustrated in Figure 6.2.

6.3 The spherical particle shell model

Table 6.1 Spherical particle shell parameters. x_0 is the value of x at the central radius for the shell. The x_0 values are determined by the values of x for which $\cos 2x = 0$. The b'_ρ value ensures that ρ_{B1} has a local $1/r$ dependence at x_0

x_0		b'_ρ
$3\pi/4$	2.36	0.269
$5\pi/4$	3.93	-0.113
$7\pi/4$	5.50	0.100
$9\pi/4$	7.07	-0.066
$11\pi/4$	8.64	0.061
$13\pi/4$	10.21	-0.047
$15\pi/4$	11.78	0.044
$17\pi/4$	13.35	-0.036

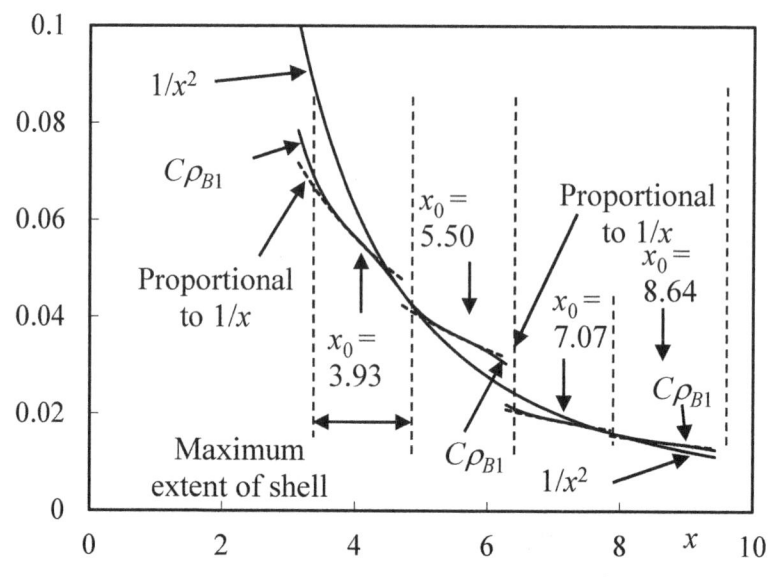

Figure 6.2 Variation of solution $C\rho_{B1}(x)$ compared to $1/x^2$ and local $1/x$ variations (shown by dashed lines). C is a constant introduced in the Book, Appendix 6A, Section 6A.4

Thus a spherical shell particle has been constructed, Figure 6.3.

The shell is bounded in r by discontinuities in ρ_B and m_B. This model can be extended to include many shells, each shell separated from the adjacent ones by discontinuities. If the shells have thickness of $\pi/2$ there is no gap between the edge of one shell and the edge of the next and the volume is completely filled from the innermost shell to the outermost shell. Details are shown in Figure 6.2. When $b_p' = 0$ the $C\rho_{B1}$ solution is shown on Figure 6.2 as $1/x^2$ where C is a constant, and this describes the relative values at the centres of the shells. When b_p' takes on the values in Table 6.1 the shells appear. The dashed lines are the $1/r$ variations in each shell. The solid lines are the actual variations for these b_p' values. Hence the requirement for $1/r$ variations for ρ_{B1} and m_B is met locally within shells in a multishell particle. Each shell is separated from the adjacent ones by discontinuities in ρ_{B1} and m_B.

6.4 The thickness of the shells

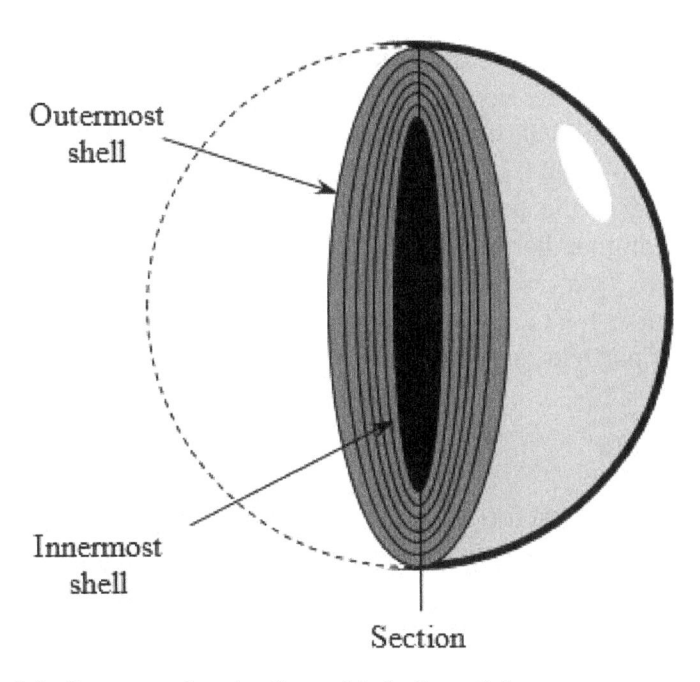

Figure 6.3 Cut away sketch of a multi-shell particle

6.4 The thickness of the shells

In Section 6A.2 in the Book it is shown from an examination of the steady state Field Equations that in each shell

$$\frac{m_A}{\rho_A} = -f \frac{c^2}{\varepsilon_0 \omega_0^2}$$

(6.6)

where m_A and ρ_A are in anti-phase and f is a real positive number. Since for particles in the subset $m_A/\rho_A \propto \omega_0^{1/2}$, see equation (6.4) above, then f is proportional to $\omega_0^{5/2}$. It is shown in the Book Section 6A.7 that

$$2\Delta x_{max} = \frac{16f^2}{x_0}$$

(6.7)

Baryons and mesons

where $2\Delta x_{max}$ is the maximum allowed filling of the shell centred on x_0 provided that this does not exceed $\pi/2$ at which the volume can be completely filled from x_0 down to the innermost shell. It is shown in the Book that there are no solutions when $f < 1$ without invoking the presence of higher harmonics. We shall conjecture that $f = 1$ is a condition for a stable particle in that this is the lowest value of f, corresponding to the lowest angular frequency without invoking the presence of higher harmonics. So let's put $f = 1$. When $2\Delta x_{max} = \pi/2$, using (6.7), $x_0 = 32/\pi = 10.2$. This happens to correspond to a value in Table 6.1. The outer radius of the spherical particle is 11.0 in x units and the particle can be filled to this radius.

6.5 The proton

In this section we develop a model for the proton and the model is used in the next section to obtain expressions for the constants A and K. It is proposed that the proton is a multi-shell particle. The rationale is as follows. In the Book detailed solutions are obtained applicable to spin ½ particles, and therefore appropriate for a proton. These solutions can be approximated as functions of r_c in cylindrical co-ordinates. Putting $r_c = r \sin\theta$ where r_c is the radius in cylindrical co-ordinates r_c, z, ϕ then the expressions in the Book, Appendix 6B, equations (6B.1) and (6B.2), for Group B spin ½ particles may be recast, so that

$$\rho_A = \sqrt{\frac{2}{\pi x}} \left[\frac{B_{\rho A}}{r_c^{1/2}} \cos x' + \frac{B_{\rho B}}{r_c^{1/2}} \sin x' \sin\theta \right] \exp\left(\pm \frac{i\phi}{2}\right)$$

(6.8)

where now $x = \omega_{0p} r_c/c$, ω_{0p} is the proton angular frequency, $x' = x - \pi/4$ and $x \gg 1$. A similar expression applies for the oscillatory gravitational mass density together with extra terms.

These solutions can be matched to those of the spherical particle over a limited range of θ by consideration of the following three points in comparing (6.8) with (6.5):

(1) $\sin x'$ and $\cos x'$ are equal to $\sin x$ and $\cos x$ used for the spherical particle. There is a shift of $\pi/4$ upwards in the values of x_0 for the proton from those of the spherical particle.
(2) (6.8) contains the extra factor $\exp(\pm i\phi/2)$. However in

6.5 The proton

forming m_B and ρ_{B1} the factor disappears.
(3) (6.8) has an additional $\sin\theta$ factor and r_c instead of r. Referring to Figure 6.4, it can be seen that there is a range $\Delta\theta$ around $\theta = \pi/2$ over which there is an overlap of the two geometrical configurations. $\sin\theta$ can be replaced by unity in the vicinity of $\theta = \pi/2$.

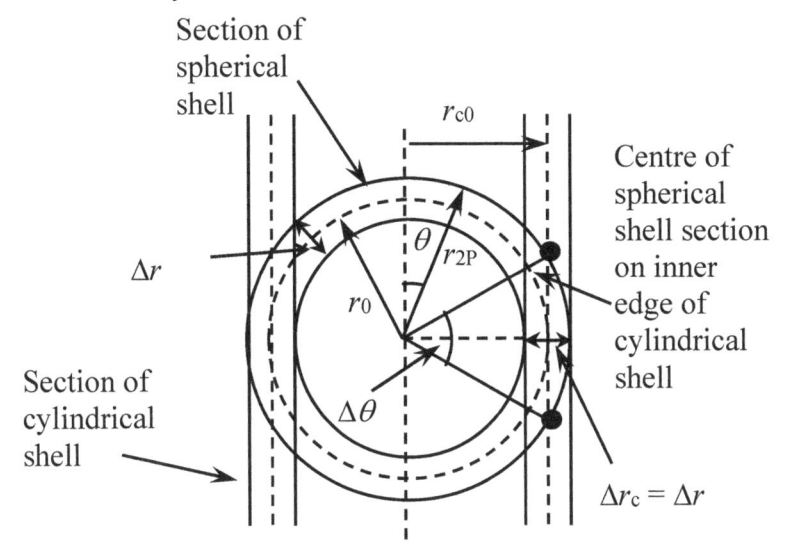

Figure 6.4 Overlap between cylindrical and spherical particle configurations

The proton outer radius is r_{2P}. The angle $\Delta\theta$ can be estimated in terms of $x_{2P} = \omega_{0P} r_{2P}/c$ and $2\Delta x_{max} = \Delta r \omega_{0P}/c$ by the condition that the departure of the outermost cylindrical shell from the spherical shell is one half of the shell thickness and so

$$\Delta\theta = 2\cos^{-1}\left(1 - \frac{|\Delta x|_{max}}{x_{2P}}\right)$$

(6.9)

Essentially the proton is a cylindrical structure. It is proposed that the proton is a multi-shell particle similar to the spherical particle at its equator, but limited to a range of θ of $\pm\Delta\theta/2$ above and below $\theta = \pi/2$, Figure 6.5.

Baryons and mesons

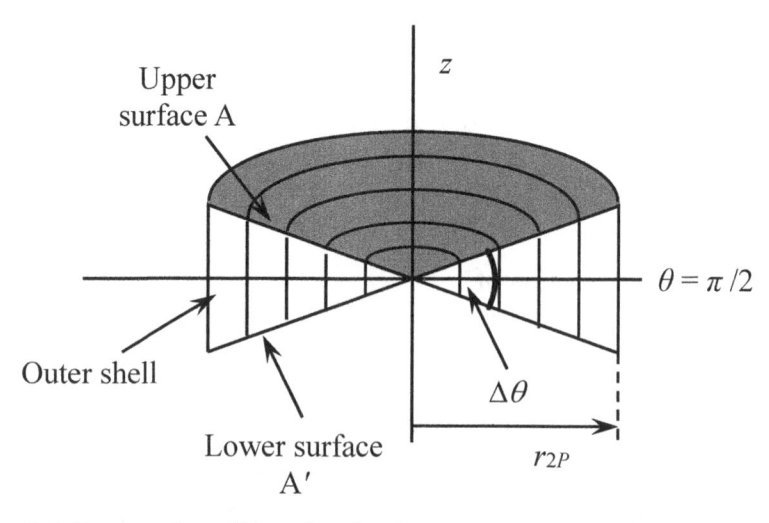

Figure 6.5 Proposed configuration for the proton

Because there is a $\pi/4$ shift in the shell structure, the shells are now centred at multiples of $\pi/2$. In the previous section it is conjectured that $f = 1$ is a condition for a stable particle. Since the proton is stable we shall choose this condition for the proton.

Using equation (6.7), when $f = 1$ and the shells are filled, $x_0 = 10.2$. The next higher proton shell is at $x_0 = 11.0$ for which $2\Delta x_{max} = 1.45$. We shall estimate the radius out to which the particle is filled by simply adding Δx_{max} to x_0 resulting in $x_{2P} = 11.7$. We can repeat this calculation where the next shell is included to obtain $x_0 = 12.6$ and $2\Delta x_{max} = 1.3$ and adding to 11.7 gives $x_{2P} = 13.0$.

6.6 Expressions for the constants A and K

The fundamental constants are (Section 4.3) are c, ε_0, A_m, A_ρ and $\hbar$. The expressions in equations (6.1) to (6.4) also contain A and K. In this section we obtain expressions for the constants A and K in terms of proton parameters. These expressions are used in Chapter 8 to systematically obtain the values of the new fundamental constants A_m and A_ρ. In order to obtain expressions for A and K the analysis proceeds by modelling the proton as an exact portion of the spherical particle restricted to $\pm \Delta\theta/2$ above and below $\theta = \pi/2$. The electric field at the surface using Gauss' theorem is

6.6 Expressions for the constants A and K

$$E = \frac{q}{2\pi\varepsilon_0 r_{2P}^2 \Delta\theta}$$

where q refers to the proton. Integrating, the potential is given by

$$m_B = \frac{q}{2\pi\varepsilon_0 r_{2P} \Delta\theta}$$

This is an approximation since these steps neglect the leakage of the electric field from the now exposed upper and lower surfaces. This is remedied in Chapter 9. We can find m_{BM} by taking m_B to be proportional to $1/r^2$ (see Figure 6.2) so that m_{BM} is three times the value at the outer surface and the proton gravitational mass is

$$m_{BM} V_P = \frac{3q}{2\pi\varepsilon_0 x_{2P} \Delta\theta} \left(\frac{\omega_{0P}}{c}\right) V_P$$

(6.10)

noting that we have found the value of x_{2P} in the previous section and where the proton volume is given by

$$V_P = \frac{2\pi x_{2P}^3 \Delta\theta}{3} \left(\frac{c^3}{\omega_{0P}^3}\right)$$

(6.11)

From (6.10) and (6.11) the constant A is given by, where $M_0 = A\omega_0$ for any particle with angular frequency ω_0,

$$A = \frac{3qV_P}{2\pi\varepsilon_0 x_{2P}\Delta\theta c} = \frac{qx_{2P}^2 c^2}{\varepsilon_0 \omega_{0P}^3}$$

(6.12)

From (6.6) with $f = 1$ for the proton and using (6.6) and (6.10),

$$K = \rho_{AM}^* m_{AM} = -\frac{3q\omega_{0P}^2}{A_m 2\pi r_{2P}\Delta\theta c^2} = -\frac{qx_{2P}^2}{V_P A_m}$$

(6.13)

The expressions for A and K are used in Chapters 7 and 8. Note that since

$$q = A_\rho \rho_{AM}^* \rho_{AM} V_P$$

then from (6.6) and (6.13),

$$\frac{A_\rho}{A_m} \left(\frac{\varepsilon_0 \omega_{0P}^2}{c^2}\right) x_{2P}^2 = 1$$

(6.14)

Baryons and mesons

This is used in Chapter 8 as an alternative way to evaluate x_{2P}. From (6.3), (6.12) and (6.13) for any particle in the subset,

$$\frac{|m_{AM}|}{|\rho_{AM}|} = \left(\frac{c^2}{\varepsilon_0 \omega_{0P}^3}\right)\left(\frac{v_P}{v}\right)\omega_0 = \left(\frac{c^2}{\varepsilon_0 \omega_{0P}^2}\right)\left(\frac{\omega_0}{\omega_{0P}}\right)^{1/2} \quad (6.15)$$

6.7 Anti-protons

We can account for the antiproton as follows. Use the proton oscillatory solutions. There are two possibilities. The first is to make the transformation $A_m = -A_m$, $A_\rho = -A_\rho$. The transformed steady state solutions satisfy the steady state Field Equations. $\rho^*_{AM} m_{AM}$ continues to be negative. $|q|$ is the same, the angular frequency is the same and the particle dimensions are the same. The new particle, the anti-proton, has a negative charge and a negative gravitational mass. Alternatively put $A_\rho = -A_\rho$ leaving A_m unchanged as a positive constant. The new particle has negative charge but retains a positive gravitational mass. The consequences of these anti-proton alternatives are discussed in Chapter 11.

6.8 Neutral particles

We now examine how neutral particles, and therefore the neutron, can arise using the multi-shell model, whether in the form of the spherical particle or the variant introduced for the proton. In order for a particle to behave as a neutral particle the charge must be near zero and the external electric potential in the far field must be near zero. The model presented here is based on there being positively and negatively charged shells or regions so that the net particle charge is essentially zero. The oscillatory solutions are the same as the charged particle, the angular frequency is the same and the particle dimensions and number of shells are the same. For any one shell consider $A_m \to \pm A_m$, $A_\rho \to \pm A_\rho$ and under the Laplace approximation the steady state Field Equations are satisfied. However because the angular frequency remains unchanged the gravitational mass must remain at M_0 or becomes $-M_0$. This reduces the possibilities to:

a neutral particle with A_m positive A_ρ positive shells alternating

6.8 Neutral particles

with A_m positive A_ρ negative shells as a function of radius

the neutral anti-particle with A_m negative A_ρ positive shells alternating with A_m negative A_ρ negative shells as a function of radius

the neutral anti-particle with A_m positive A_ρ positive shells alternating with A_m positive A_ρ negative shells as a function of radius

However in order for the electric potential in the far field to be near zero, then m_B must drop to near zero at the particle boundary in a similar manner to ρ_{B1}. If the modulus of the charge per shell is approximately the same for adjacent shells, then over all shells the net charge can be zero. The shell thicknesses can adjust to make this exact.

Thus the model for the neutron is based on the proton with alternating sign of the charge density in successive shells, depicted in Figure 6.6. The proton oscillatory solutions apply, the same dimensions result and the inertial mass and gravitational mass are approximately the same as those of the proton.

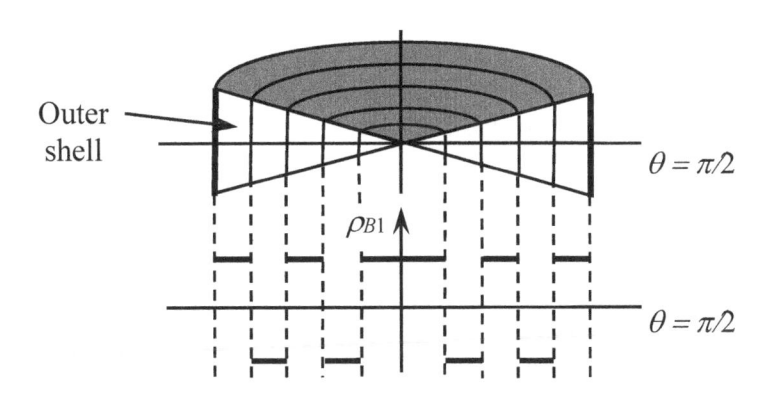

Figure 6.6 The proposed configuration for the neutron

The negative charge in the A_ρ negative shells can circulate in the same direction as the positive charge in the A_ρ positive shells, but not necessarily so – the theory has not been developed to determine the direction of circulating currents. Hence the magnetic moment may range from contributions from all shells adding to overall cancellation

Baryons and mesons

over the shells. Similarly for the proton the charge may circulate in different directions from shell to shell. Hence there is a range of theoretical magnetic moments for both particles.

At the surface the steady state gravitational density must be equal to or greater than the oscillatory gravitational mass density amplitude in order to ensure compliance with the relativistic constraint (see Chapter 2). Similarly at the surface the steady state charge density must be equal to or greater than the oscillatory charge density amplitude. Suppose that when neutrons condensed out both these conditions were satisfied, but one of them only just. There are two possibilities: Case 1, $\rho_B > |\rho_A|$, $m_B = |m_A|$. Hence $1/\rho_B < A_\rho$ and $1/m_B = A_m$; Case 2, $m_B > |m_A|$, $\rho_B = |\rho_A|$. Hence $1/m_B < A_m$ and $1/\rho_B = A_\rho$. It is shown in Chapter 8 that Case 2 applies. Since $\rho_B/|\rho_A|$ falls with angular frequency, see equation (6.2), then ρ_B becomes less than $|\rho_A|$ at frequencies above that for the proton/neutron and so higher frequency particles are predicted to be unstable.

6.9 Mesons

Kaons. So we have set up a spherical shell model where the particle spin is zero and the first application is with particles with spin one half, i.e. protons and neutrons. We now remedy this, but only partially. The spherical shell model with spin zero can be applied to the kaons, resulting in various two shell configurations. However the details on how this is achieved have to wait until the introduction of composite particles in Chapter 14.

Pions. We now turn attention to the other solutions in Group A from Chapter 2, Section 2.4, where the spin is zero but instead of spherical solutions for the oscillatory fields, they are of the form $\cos\theta/r$ and so the steady state gravitational mass density and charge density are proportional to $\cos^2\theta/r^2$.

The steady state solutions are constant when $|\cos\theta| = ar$ where a is a constant and this is the case on the 'solution equipotential' circles shown in Figure 6.7.

6.9 Mesons

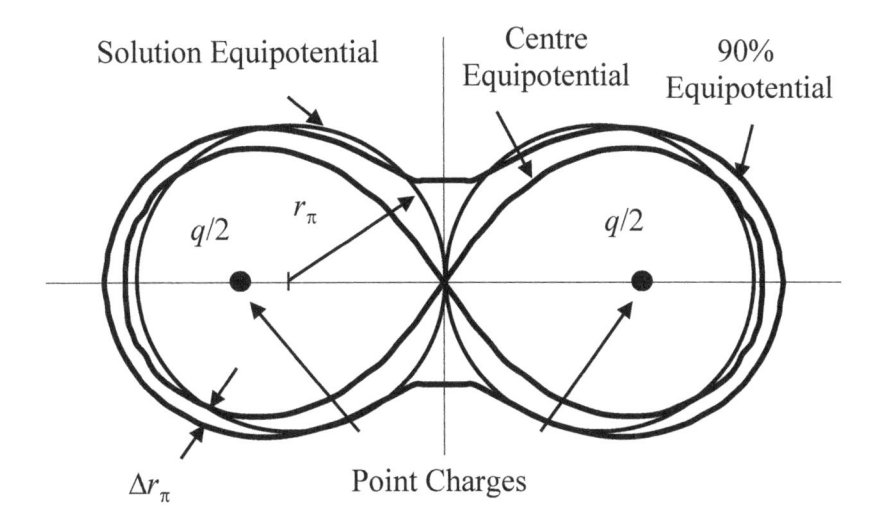

Figure 6.7 Comparison of the solution equipotential and the equipotentials for the pion point charge model

So the particle, consisting of two spherical shells touching at the centre based on two such circles rotated about the z axis, will have two shells of constant steady state values. Now replace the charged shells by the two point charges, each $q/2$, but shifted along the z axis from the circle centres as shown in Figure 6.7. Equipotentials for the two charge system are sketched. One passes through the centre and the other has a potential of 90% of the centre equipotential. Since these describe the potentials from point charges they represent Laplace solutions. The volume between the centre and 90% equipotentials is a region where the two point charge system equipotentials and the locus for constant steady state gravitational mass density roughly coincide.

We can now use this configuration to set up a model for a charged pion. We take the spherical circumference to be one free space wavelength and so the radius x_π in units of $\omega_0 r_\pi/c$ is 1 and the spherical shell thickness $\Delta x_\pi = 0.1$. The volume is given by, using (6.3),

$$8\pi x_\pi^2 \Delta x_\pi \left(\frac{c^3}{\omega_{0\pi}^3}\right) = V_P \left(\frac{\omega_{0\pi}}{\omega_{0P}}\right)^{1/2}$$

and so

$$\omega_{0\pi} = \left(\frac{8\pi x_\pi^2 \Delta x_\pi c^3 \omega_{0P}^{1/2}}{v_P}\right)^{2/7}$$

We have found the angular frequency from which we can obtain the pion energy – the calculation is in Chapter 8.

A model for the neutral pion is proposed in Section 12.6 in the Book. An objection to this model is that it will have an electric dipole moment, which is not observed experimentally. An alternative is to surround the thin spherical shells containing one charge with thin shells of the opposite charge. Since the internal charge density oscillatory solutions are the same as in the charged pions, then the rest energy is predicted to be the same as the charged pions.

6.10 Summary

The aim of this chapter is to propose models for the proton, neutron, kaon and pion. The first step has been to investigate a spherical particle model. A Laplace approximation to Poisson's equation is used. The problem that the steady state solutions derived from the oscillatory solutions are proportional to $1/r^2$ and therefore do not satisfy Laplace's equation is solved by introducing local $1/r$ variations separated by discontinuities. This leads to a multi-shell model. The maximum thickness of the shells depends on the outer x_0 radius of the particle and a parameter f related to the angular frequency. A model is proposed for the proton based on an adaptation of the multi-shell spherical model. The expression for $\hbar$ obtained in the previous chapter contains the constants A and K. Expressions are obtained in this chapter for A and K in terms of proton parameters. Neutral particles are predicted from the theory as a consequence of the Laplace approximation used in the multi-shell model. Outline models are proposed for the kaon and pion. Anti-particles are accounted for in the scheme.

Chapter 7

Leptons

7.1 Introduction

In the previous chapter, in this chapter and in the next chapter, we apply the general particle scheme to particular particles and show that theoretical expressions for various particle parameters can be obtained. In the previous chapter baryons and mesons are investigated. Attention is given in this chapter to leptons, particularly to the electron and the muon together with their anti-particles.

The electron is the particle whose properties are dominant in atomic physics, molecular physics and solid state physics. These are the areas where conventional quantum mechanics has overwhelming success and provides the understanding for the exploitation of electrons in modern applied physics and engineering.

Here we are concerned with models for the electron and muon, their anti-particles and their neutral counterparts, the neutrinos, which lead to predictions of their basic observed properties. In the previous chapter the proton is at an angular frequency with $f = 1$ which is the lowest for a stable particle. So how do we describe the electron which we know experimentally has a much lower inertial mass? The key step is to construct standing wave type solutions of the Field Equations using the Group B solutions from the previous chapter. This results in a set of equations with properties distinct from those of the spherical particle and lead to models for the electron and the muon. These equations also have the appearance of those for a very low angular frequency particle and this is the basis for the proposed mechanism to account for neutrinos. The tau is the third lepton and we do not yet have a suggestion for its structure.

7.2 Standing wave solutions

Our starting point is with the Group B set of solutions with spin ½, see the Book, Chapter 2, Section 2.4 and Chapter 6, Section 6.4. The

Leptons

oscillatory solutions have terms which are independent of z (well, nearly so, the detail is in the Book), and this allows us to take the next step which is special to leptons. We adapt the solution scheme to the case where m_A and ρ_A are composed of a standing wave along the z axis. Thus put

$$m_A = 2m_{Ak} \cos\{k(z - z_0)\} \exp i\omega t$$
$$\rho_A = 2\rho_{Ak} \cos\{k(z - z_0)\} \exp i\omega t$$

The factor of 2 is to maintain consistency with the analysis in the Book. Substituting each component into the scalar oscillatory Field Equations, and dropping inconvenient sine terms (justified by the Book analysis), they can be recast as

$$\frac{\partial^2 m_{Ak}}{\partial x^2} + \frac{\partial^2 m_{Ak}}{\partial y^2} + \frac{\partial^2 m_{Ak}}{\partial z^2} + \frac{\omega_{00}^2}{c^2} m_{Ak} = -\frac{\rho_{Ak}}{\varepsilon_0}$$

$$\frac{\partial^2 \rho_{Ak}}{\partial x^2} + \frac{\partial^2 \rho_{Ak}}{\partial y^2} + \frac{\partial^2 \rho_{Ak}}{\partial z^2} + \frac{\omega_{00}^2}{c^2} \rho_{Ak} = \frac{m_{Ak}}{\varepsilon_0}$$

(7.1)

where

$$\frac{\omega_{00}^2}{c^2} = \frac{\omega_0^2}{c^2} - k^2$$

The solutions of (7.1) for m_{Ak} and ρ_{Ak} are given by the expressions in the Group B equations (6B.1) and (6B.2) in the Book where x is now $\omega_{00} r/c$, with ω_{00} replacing ω_0. These are the solutions used in the following models for the electron and muon with ω_{00} much less than ω_0 and $k \cong \omega_0/c$. There is a major consequence of replacing ω_0 by ω_{00} in the solutions. These solutions without the multiplying standing wave factor can be those for a low inertial mass particle, and can account for the existence of neutrinos, see section 7.6. The solutions contain perturbations, and perturbations of perturbations and so on, and this series converges if $\omega^2_{00} > c^2/\varepsilon_0$. This condition is important for neutrinos.

7.3 A model for the electron

Appendix 7A in the Book shows that the oscillatory solutions may be

7.3 A model for the electron

recast as follows:

$$\rho_{Ak} = \left[\frac{B_{\rho A}}{r_c^{1/2}} + \frac{B_{\rho B}}{2}\left(\frac{\omega_{00}}{c}\right)r_c^{1/2} + C_{\rho c}r_c^{1/2}\right]\exp\left(\pm\frac{i\phi}{2}\right)$$

$$m_{Ak} = \left[\frac{B_{mA1}}{r_c^{1/2}} - \frac{B_{\rho A}b^2}{4\varepsilon_0 r_c^{1/2}}\right]\exp\left(\pm\frac{i\phi}{2}\right)$$

where $B_{\rho A}$, $B_{\rho B}$, $C_{\rho c}$ and B_{mA1} are constants, $b^2 = \Delta r_c^2 + z^2$ and $\Delta r_c = r - r_{c0}$. It can be seen that ρ_{Ak} is independent of z. By consideration of the steady state Field Equation in cylindrical coordinates,

$$\frac{1}{r_c}\frac{\partial}{\partial r_c}\left(r_c\frac{\partial m_B}{\partial r_c}\right) + \frac{\partial^2 m_B}{\partial z^2} = -\frac{\rho_B}{\varepsilon_0}$$

and with b much less than r_{c0}, it turns out that b is a constant given by

$$b^2 = -4\varepsilon_0 A_{\rho e}/A_{me}$$

(7.2)

where $A_{\rho e}$ and A_{me} are A_ρ and A_m for the electron containing the signs of the steady state values. Equation (7.2) requires that for the electron to have a positive gravitational mass, the charge has to be negative. The resulting structure is shown in Figure 7.1. The particle is localised on circles in the r_c, z planes whose centres lie on the circle $z = 0$, $r_c = r_{c0}$ in the r_c, ϕ plane. Δr_{c0} and Δz_0 are central values and the central value of the radius b of the local circle is given by $b^2 = \Delta r_{c0}^2 = \Delta z_0^2$. Hence the proposed configuration is a hollow torus of radius r_{c0} to the centre of a circle of radius b. From equation (7.2) we can find the radius of the local circles once we know A_ρ and A_m - we do in Chapter 8. The steady state gravitational mass density is discontinuous at the boundary – the internal charge is negative, the internal gravitational mass is positive but the external electric potential is required to be negative.

Leptons

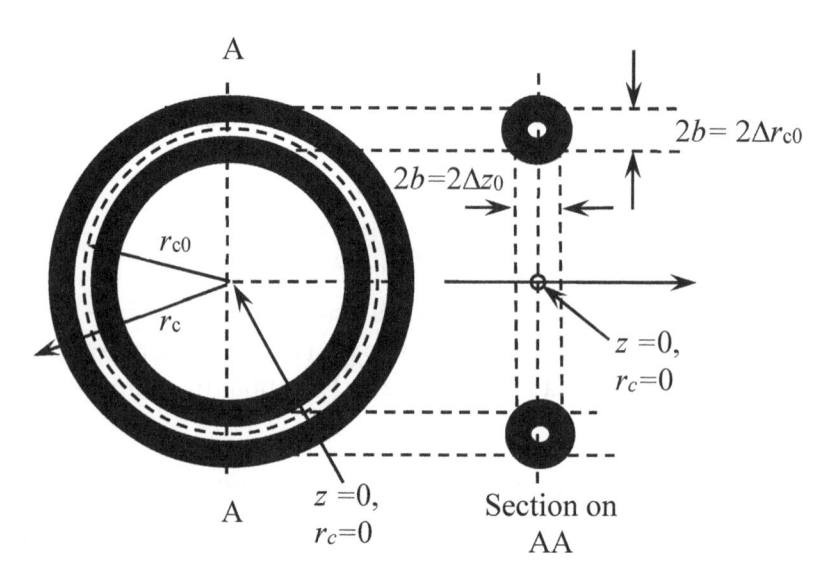

Figure 7.1 Proposed structure for the electron

It could be the case that there is a range of b values, resulting in a thick wall torus as shown in Figure 7.1. However a single value of b is implied by (7.2). The particle then has a thin wall, and we can put

$$\frac{m_{AM}}{\rho_{AM}} = \frac{m_{Ak}}{\rho_{Ak}}$$

It is shown in appendix 7A in the Book that equation (6.4) in Section 6.2 applies to the electron and that

$$\frac{m_{AM}}{\rho_{AM}} = -\frac{4A_\rho}{A_m} = -\left(\frac{AA_\rho \omega_{0e}}{qA_m}\right)^{1/2}$$

(7.3)

where ω_{0e} is the electron angular frequency.

The detailed treatment of the external oscillatory field has been dealt with in Chapter 3. Treat the electron torus as a circular wire-like object. A circular loop of wire has resonances for frequencies at which the circumference is an integer number of free space wavelengths. If the circumference of the torus is one wavelength in free space, the oscillatory electric potential around the circumference can be expressed as a superposition with components at ω_{0e}, $2\omega_{0e}$, $3\omega_{0e}$ etc. in order to match the internal oscillatory gravitational mass density

7.5 The muon

variation with ϕ. Hence we put $r_{c0} = c/\omega_{0e}$. Thus we have the major parameters for the electron structure, the torus radius and the cross-section radius.

7.4 The positron

By changing the signs of A_ρ and A_m the solution scheme is converted to that of the anti-particle, the positron, with positive charge and negative gravitational mass. It is pointed out in in the previous section, having A_ρ and A_m of the same sign is not allowed, and so it is not possible to form multiple negatively and positively charged regions in either the electron or the positron with either the gravitational mass density always positive or always negative. This precludes a neutral particle being formed with a gravitational mass of similar magnitude as the charged particle and therefore of similar inertial mass. A different mechanism for a neutral counterpart is required, discussed in the section on neutrinos below.

7.5 The muon

In this section a structure is proposed for the muon and is used to obtain an expression for its angular frequency. We shall assume that we can obtain solutions for m_{Ak} and ρ_{Ak} which are not dependent on z. The independence from z invites examination of a cylindrical configuration, Figure 7.2.

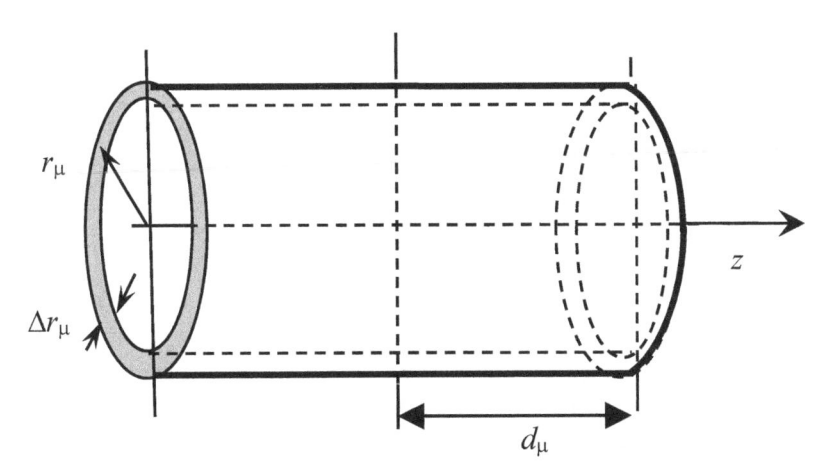

Figure 7.2 Proposed structure for the muon

Leptons

The radius is taken to be $r_\mu = c/\omega_{0\mu}$ corresponding to a circumference of one free space wavelength as adopted for the electron (i.e. a radius of unity in x units) and the length is taken to be $\pi c/\omega_{0\mu}$ corresponding to a half wavelength standing wave along z (on Figure 7.2 $d_\mu = \pi/2$ in x units).

We take the gravitational mass density to be a Laplace solution and so we do not require continuity of the internal and external gravitational mass densities. This is discussed in Section 2.5 in Chapter 2 and results in the calculated value for the muon energy in Chapter 8 to be higher than that obtained in the Book. We require to find a charge system which has a similar potential distribution to the steady state solutions. This problem was also tackled with the pion in Chapter 6, Section 6.9. We adopt for comparison the two charge system potentials as in the pion case – a cylinder of length π in x units can be accommodated and the thickness Δx_μ is 0.1. Equating the volume to the expression in equation (6.3) in Section 6.2, we have

$$0.2\pi^2 \left(\frac{c^3}{\omega_{0\mu}^3}\right) = V_P \left(\frac{\omega_{0\mu}}{\omega_{0P}}\right)^{1/2}$$

and

$$\omega_{0\mu} = \left(\frac{0.2\pi^2 c^3 \omega_{0P}^{1/2}}{V_P}\right)^{2/7}$$

The analysis applies to the anti-muon by changing the signs of A_m and A_ρ and the oscillatory solutions remain unchanged.

7.6 Neutrinos

The feature common to the lepton models above is the use of the standing wave construction and this is exploited in the model advanced here to explain the existence of the neutral lepton counterparts. If ω_{00} is replaced by a new ω_0, then the form (7.1) of the oscillatory Field Equations solved for the leptons is the form proposed to be solved for the neutrinos without a k vector. Therefore the solution of the oscillatory Field Equations for each lepton generates a solution for the corresponding neutrino. Since ω_{00} is a very low angular frequency

7.7 Summary

compared to the lepton angular frequency, then the neutrino counterpart has a very low inertial mass. This implies a very low gravitational mass. This can be achieved by having A_m of different signs in two coupled tori resulting in near cancellation of the gravitational mass

Appendix 7C in the Book examines this model for the electron neutrino. It is concluded that the electron neutrino consists of a superposition of negatively and positively charged tori. The steady state solutions of the corresponding lepton apply (except for a $\cos^2\{k(z - z_0)\}$ factor) with a change in sign of A_m accompanied by a corresponding change in sign of A_ρ resulting in either exact or near cancellation of the electric charge. The tori have similar radii to the electron torus. The particle has very low gravitational and inertial masses. The angular frequency has not been determined. For stability we require that $c^2/\varepsilon_0 \omega^2{}_{00} < 1$ and so the angular frequency must be larger than $\varepsilon_0^{-1/2} c$.

The discontinuity in the gravitational mass density at the electron boundary, the possibility that the positron has a positive gravitational mass and why should the neutrinos be composed of two tori rather than one as in the electron and positron – these issues are resolved in Chapter 17 where the structure of particle boundaries is discussed.

7.7 Summary

The aim of this chapter is to propose models for the electron and muon, their antiparticles and their neutral counterparts, the neutrinos. The key step is to construct standing wave type solutions of the oscillatory Field Equations. This construction is applied to Group B spin ½ oscillatory solutions and a low angular frequency (relative to the proton) particle, the electron, is modelled.

A hollow torus configuration is proposed for the electron and positron. The circumference of the centre circle of the torus is one free space wavelength. A hollow cylinder model is proposed for the muon and anti-muon. The circumference of the cylinder is one free space wavelength. An expression is obtained for the angular frequency of the muon relative to the proton.

There are two significant consequences of the standing wave construction. First neutral counterparts with similar inertial masses to the charged particles are not possible. Secondly there can exist neutral

Leptons

particles, the neutrinos, with very low angular frequencies and hence very low inertial masses. The oscillatory Field Equations have the appearance of those for a very low angular frequency particle and this is proposed as the mechanism to account for neutrinos. It is proposed that the electron neutrino consists of a superposition of negatively and positively charged tori, with torus radii similar to that of the electron torus.

We conclude that single omega solutions are possible for the leptons: the electron, muon, electron neutrino, muon neutrino and their anti-particles.

Chapter 8

Constants and particle parameters

8.1 Introduction

In this and the preceding two chapters we apply the general particle scheme of Chapter 2 to particular particles and show that theoretical expressions for various particle parameters can be obtained. In Chapter 6 baryons and mesons are investigated. Chapter 7 proposes models for the electron and muon, their anti-particles and their neutral counterparts, the neutrinos. Building on the results from these chapters, specific aims of this chapter are to obtain the values of the electronic charge and the gravitational constant, and to calculate various particle parameters from the five fundamental constants.

The chapter proceeds by setting out the numerical values of the fundamental constants. The values of A_m and A_ρ are initially unknown and the values of two other constants have to be assumed. The ones chosen are the proton and electron energies. The next step is to establish the values of the proton structure constants, x_{2P} and $\Delta\theta$. The scene is then set to calculate the theoretical values of the electronic charge, the fine structure constant and the gravitational constant. The predicted energies of the muon, pion and kaon are calculated and collated with other particle energy data. The predicted structures for the various particles are brought together pictorially in Figure 8.1.

8.2 Fundamental constants

The first step is to be clear as to the values of the fundamental constants and this is dealt with in this section. The formalism requires five fundamental constants, the special velocity magnitude c, Planck's constant $h = 2\pi\hbar$, $\alpha = 1/\varepsilon_0$ where ε_0 is the permittivity of free space, and the steady state constants A_m and A_ρ. Values here and throughout this chapter are based on those provided by the Particle Data Group (2014),

Constants and particle parameters

$c = 2.99792458 \times 10^8$ ms⁻¹ (exact)

$\varepsilon_0 = 8.854187817 \ldots \times 10^{-12}$ Farad m⁻¹ (derived exactly from c and the permeability of free space $\mu_0 = 4\pi \times 10^{-7}$ N A⁻² exact) and hence $\alpha = 1/\varepsilon_0 = 1.129 \times 10^{11}$ m⁻²

$h = 6.62606957 \times 10^{-34}$ Js and hence

$\hbar = h/2\pi = 1.054571726 \times 10^{-34}$ Js

Note that this implies that the farad has dimensions of m³. Note also that the values for c and ε_0 are exact – they have these values by definition not measurement. The values of A_m and A_ρ are initially unknown and the first step is to derive them from the formalism. In order to do so the values of two other constants have to be assumed. The ones chosen are the energies of the proton and the electron, with the following rounded values, $W_P = 938.3$ MeV and $W_e = 0.511$ MeV. These are converted from MeV to joules using

$$1eV = 1.602176565 \times 10^{-19} \text{ J}$$

So far as the formalism is concerned the conversion factor is arbitrary - the fact that it is based on the measured value of the electronic charge does not jeopardise the integrity of the theoretical prediction of the electronic charge which is made below. Equating these energies to $\hbar\omega_{0P}$ and $\hbar\omega_{0e}$ respectively results in

$\omega_{0P} = 1.43 \times 10^{24}$ rad.s⁻¹ $\qquad \omega_{0e} = 7.76 \times 10^{20}$ rad.s⁻¹

Useful parameters are

$\omega_{0P}/\omega_{0e} = 1836 \qquad c/\omega_{0P} = 2.10 \times 10^{-16}$ m

$c/\omega_{0e} = 3.86 \times 10^{-13}$ m $\qquad \varepsilon_0 \omega_{0P}^2/c^2 = 2.00 \times 10^{20}$

The contention is that all other physical parameters are expressible in terms of c, $\hbar$, ε_0, A_m and A_ρ and the set used initially is c, $\hbar$, ε_0, ω_{0P} and ω_{0e}. The major parameters calculated below are the values of the electronic charge, the fine structure constant, A_m, A_ρ, A, K and the gravitational constant.

Many expressions contain the parameters x_{2P} and $\Delta\theta$ which relate to the structure of the proton. From (7.3) for the electron,

8.3 The electronic charge

$$\frac{|m_{AM}|}{|p_{AM}|} = \frac{4A_\rho}{A_m}$$

and from (6.4) and (6.6) for the electron

$$\frac{|m_{AM}|}{|p_{AM}|} = \left(\frac{c^2}{\varepsilon_0 \omega_{0P}^2}\right) \frac{\omega_{0e}^{1/2}}{\omega_{0P}^{1/2}}$$

and using (6.14)

$$x_{2P} = 2\left(\frac{\omega_{0P}}{\omega_{0e}}\right)^{1/4}$$

(8.1)

giving $x_{2P} = 13.1$. Now

$$x_{2P} = x_0 + \pi/4 + |\Delta x|_{max}$$

where x_0 is the shell centre for the spherical particle. From Chapter 6, Section 6.5, $x_{2P} = 11.7$ with almost filled shells, or with an extra nearly filled shell $x_{2P} = 13.0$. Taking all these results into account, we shall take a rounded value and put $x_{2P} = 13$ and assume that the proton shells are filled to this outer radius in x units. Using (6.9), $\Delta\theta = 0.70$ rad.

8.3 The electronic charge

In this section we calculate theoretical values of the magnitude of the electronic charge and the fine structure constant. From (4.6),

$$\hbar = -AA_m K$$

and using the expressions for A and K, equations (6.12) and (6.13) respectively, the modulus of the electronic charge is given by

$$e = \left(\frac{2\pi\Delta\theta\varepsilon_0 c\hbar}{3x_{2P}}\right)^{1/2}$$

(8.2)

Substituting $x_{2P} = 13$ and $\Delta\theta = 0.70$ rad and the values for $\hbar$, ε_0 and c from above gives $e = 1.8 \times 10^{-19}$ C in close agreement with the empirical value of $1.602176565 \times 10^{-19}$ C.

The fine structure constant is defined as $e^2/4\pi\varepsilon_0\hbar c$ where the

Constants and particle parameters

factor $1/4\pi\varepsilon_0$ is required in SI units. From (8.2)

$$\frac{e^2}{4\pi\varepsilon_0 \hbar c} = \frac{\Delta\theta}{6x_{2P}} = \frac{1}{112}$$

which is close to the empirical value of around 1/137. Note that the expression for the fine structure constant contains only a numerical factor and the dimensionless quantities x_{2P} and $\Delta\theta$ which are related to the structure of the proton.

8.4 The volume of the proton

Using the expression from (6.11), the volume of the proton is estimated to be

$$V_P = \frac{2\pi x_{2P}^3 \Delta\theta}{3}\left(\frac{c}{\omega_{OP}}\right)^3 = 3.0 \times 10^{-44} \text{ m}^3$$

(8.3)

An equivalent sphere with the same volume has a radius of 1.9×10^{-15} m. This is a factor of over twice the measured effective radius of the proton of 0.8×10^{-15} m to 0.9×10^{-15} m.

8.5 The constants A_m and A_ρ

Using (6.14),

$$\frac{A_\rho}{A_m} = \frac{1}{x_{2P}^2}\left(\frac{c^2}{\varepsilon_0 \omega_{OP}^2}\right) = 3.0 \times 10^{-23}$$

(8.4)

Consider the values of the steady state gravitational mass and charge densities at the outer surface of the proton, from section 6.5,

$$m_B = \frac{q\omega_{OP}}{2\pi\varepsilon_0 x_{2P}\Delta\theta c} = 1.7 \times 10^6 \text{ Cm}^{-3}$$

$$\rho_B = \frac{q}{3V_P} = 2.0 \times 10^{24} \text{ Cm}^{-3}$$

In Section 6.8 there is discussion on the need to ensure that the proton steady state field magnitudes are greater than, or at least equal to, the oscillatory field amplitudes. This is to be tested at the particle outer

8.7 The gravitational constant

boundary where they are the least. Two cases are identified. Case 1, section 6.8: Put $A_m = 1/m_B = 6.0 \times 10^{-7}$ C^{-1}m^3 and hence $A_\rho = 1.8 \times 10^{-29}$ C^{-1}m^3 which is less than $1/\rho_B = 5.0 \times 10^{-25}$ C^{-1}m^3. This is unacceptable. Case 2: Put

$$A_\rho = \frac{1}{\rho_B} = \frac{3V_P}{q} \tag{8.5}$$

and $A_\rho = 5.0 \times 10^{-25}$ C^{-1}m^3 and hence $A_m = 1.7 \times 10^{-2}$ C^{-1}m^3 which is greater than $1/m_B$ and so Case 2 applies. This is the proton configuration with the minimum charge. Hence we have $A_m = 1.7 \times 10^{-2}$ C^{-1}m^3 and $A_\rho = 5.0 \times 10^{-25}$ C^{-1}m^3.

The formalism is based on the five fundamental constants c, $\hbar$, ε_0, A_m and A_ρ. If the values A_m and A_ρ had been given initially, the values of ω_{0P} and ω_{0e} could then be derived as follows. (8.1), (8.2), (8.3), (8.4) and (8.5) above and (6.9) can be solved for ω_{0P}, x_{2P}, ω_{0e}, e, V_P and $\Delta\theta$, giving the values in the previous sections.

8.6 The constants A and K

The steady state gravitational mass is given by $|M_0| = A\omega_0$ where using equation (6.10),

$$A = \frac{3qV_P}{2\pi\varepsilon_0 x_{2P}\Delta\theta c} = 1.05 \times 10^{-61} \text{ Cs rad}^{-1}$$

The constant K is given by (6.13) as

$$K = -\frac{qx_{2P}^2}{V_P A_m} = -5.87 \times 10^{28} \text{ C}^2 \text{ m}^{-6}$$

8.7 The gravitational constant

Consider two identical particles of inertial mass $\mathcal{M}_0$ and gravitational mass M_0 separated by a distance r. The attractive force between them may be written as $M_0^2/4\pi\varepsilon_0 r^2$ or, using Newton's law of gravitation, as $\mathcal{G}\mathcal{M}_0^2/r^2$ where $\mathcal{G}$ is the gravitational constant. Equating the two expressions results in

Constants and particle parameters

$$G = \frac{1}{4\pi\varepsilon_0}\left(\frac{M_0}{\mathcal{M}_0}\right)^2$$

Since M_0 is proportional to $\mathcal{M}_0$ then G is a constant as it should be and is given by

$$G = \frac{1}{4\pi\varepsilon_0}\left(\frac{Ac^2}{\hbar}\right)^2$$

as obtained previously in Chapter 4, Section 4.4. Substituting the values of the various parameters the predicted value is 7.2×10^{-11} m³ kg⁻¹ s⁻² compared with the measured value of 6.67384×10^{-11} m³ kg⁻¹ s⁻².

8.8 Particle structures and parameters

Proton/neutron. The proposed configuration is a section of the multi-shell sphere with θ restricted to an angle $\Delta\theta/2$ above and below $\theta = \pi/2$. With $x_{2P} = 13$ the radius r_{2P} is 2.7×10^{-15}m and $\Delta\theta = 0.70$ rad corresponds to 40°.

Pion. This particle is proposed to consist of two spherical shells joined at the centre of the particle. The original calculation based on continuity of the internal and gravitational mass densities and an estimate of the shell thickness resulted in a predicted rest energy of 35 MeV (see Chapter 8 of the Book). In Chapter 6, Section 6.9, of this book a new analysis gives

$$\omega_{0\pi} = \left(\frac{8\pi x_\pi^2 \Delta x_\pi c^3 \omega_{0P}^{1/2}}{V_P}\right)^{2/7}$$

Hence with $x_\pi = 1$ and $\Delta x_\pi = 0.1 x_\pi$, the predicted energy of the charged and neutral pions is 132 MeV compared to the measured values of 139.6 MeV for the charged pions and the 135.0 MeV for the neutral pion. The radius is of the spheres is predicted to be 1.5 fm and the thickness to be 0.15 fm.

Electron/positron. A torus model is proposed. The circumference is a free space electromagnetic wavelength at the electron angular frequency giving a torus large radius of 385 fm. The torus small radius, b is, from (7.2),

8.8 Particle structures and parameters

$$b = 2\left(\frac{\epsilon_0 A_\rho}{A_m}\right)^{1/2} = 3.2 \times 10^{-17} \text{ m}$$

The electron has a circular loop of steady current and the magnetic moment is given by $\mu_e = \mu_0 i S$ where i is the current and S is the area of the loop. Since $r_{co} = c/\omega_{0e}$

$$\mu_e = \frac{\mu_0 e c^2}{2\omega_{0e}} = \frac{\mu_0 e \hbar}{2M_{0e}}$$

with a value of 1.17×10^{-29} joule.m/amp (using $e = 1.602 \times 10^{-19}$ C) which is in agreement with measurement. Very accurate agreement with experiment is provided by refinements introduced by quantum electrodynamics discussed in Chapter 13.

Muon. The proposed configuration is a thin walled cylinder. The cylinder circumference is one free space electromagnetic wavelength, the thickness is 0.1 in x units and the length is π in x units. From Chapter 7, Section 7.5,

$$\omega_{0\mu} = \left(\frac{0.2\pi^2 c^3 \omega_{0P}^{1/2}}{V_P}\right)^{2/7}$$

and the muon energy is 114 MeV compared with the measured value of 105.7 MeV. The radius is 1.7 fm, the thickness 0.17 fm and the length 5.4 fm.

Neutrinos From Section 7.6 the angular frequency cannot be less than $c/\epsilon_0^{1/2}$ and this corresponds to 0.07 eV.

Particle spins, charges, inertial masses and gravitational masses. Table 8.1 collates the theoretical values of various properties for the particles discussed in Chapters 6 and 7 and compares the theoretical values with the empirical values. The spin angular momentum of the particle is proportional to the quantum number $|s|$. Table 8.1 lists these numbers. The theoretical and experimental values agree because this number has been used as a prime identifier for each particle. The values of the particle electric charges are expressed in units of e. The theoretical and experimental values agree because this number has also been used as a prime identifier for each particle.

Constants and particle parameters

Table 8.1 Comparison of predicted and observed particle properties

Particle	Symbol	Charge e	Spin	Inertial mass x c^2 MeV Exp't [1]	Predicted
Proton	p	+1	1/2	938.3	938.3 [2]
Neutron	n	0	1/2	939.6	938.3
Pion	$\pi^\pm$	±1	0	139.6	132
	π^0	0	0	135.0	132
Electron	e⁻	-1	1/2	0.511	0.511 [2]
Positron	e⁺	+1	1/2	0.511	0.511
Muon	μ^-	-1	1/2	105.7	114
Electron neutrino	ν_e	0	1/2	$< 2 \times 10^{-6}$	$\omega_{00e} > \varepsilon_0^{-1/2} c \equiv 7 \times 10^{-8}$
Muon neutrino	ν_μ	0	1/2		$\omega_{00\mu} > \varepsilon_0^{-1/2} c \equiv 7 \times 10^{-8}$

(1) Data from Particle Data Group (2014)
(2) Assumed value as part of the scheme for setting up the fundamental constants

The angular frequencies of the anti-particles and the neutral baryons and mesons are given by the same expressions as for the charged particles. All inertial masses are positive, whether particles or anti-particles, charged or neutral. The values for the proton and the electron have been assumed in setting up the system of fundamental constants. The angular frequencies of the neutrinos have not been obtained but they are greater than the values shown, see sections 7.2 and 7.6. The measured values of inertial mass times c^2 are taken from the Particle Data Group (2014). The theoretical values for the pion and muon are fairly close to the experimental values.

Geometrical configurations. Figure 8.1 presents sketches of the various particles showing the predicted dimensions. The particles are shown with the z axis up the page, except for the electron/positron.

8.9 Summary

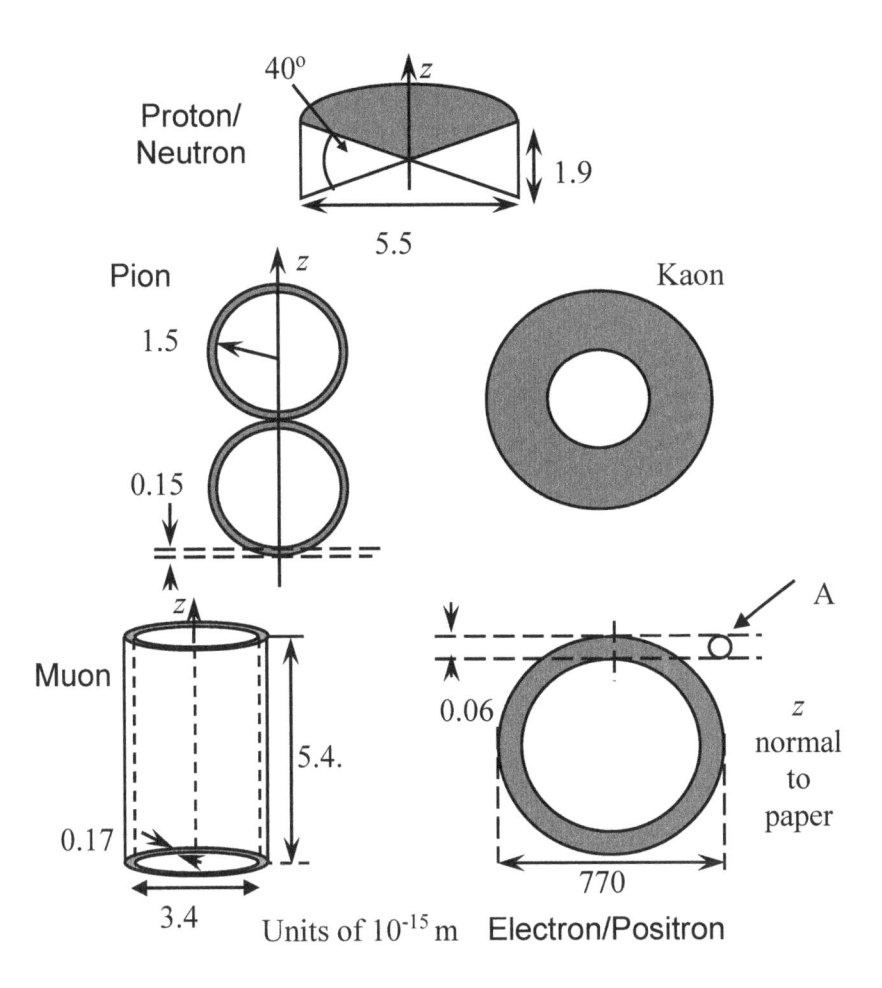

Figure 8.1 Proposed particle configurations (not to scale). The proton and neutron structures have a triangular cross section, with a 40° angle at the centre, and the structure volume is generated by rotating the triangle about the z axis. The pion has the cross section shown, and the particle volume is generated by rotating the cross section about the z axis to create two spheres touching at the centre. The kaon is shown with filled spherical shells. The muon is a hollow cylinder. The electron and positron structure is a ring with an annular cross section shown at A. The dimensions are in units of 1 fm (10^{-15} m)

8.9 Summary

Constants and particle parameters

The scheme of fundamental constants adopts the numerical values of the fundamental constants c, Planck's constant and ε_0 the permittivity of free space. The values of the remaining fundamental constants A_m and A_ρ are initially unknown. The values of the proton and electron inertial masses are assumed and the values of A_m and A_ρ are derived from them. The predicted values for the electronic charge and the gravitational mass are close to the measured values. Theoretical values for the spin angular momentum quantum number and inertial masses are tabulated for some baryons, mesons and leptons. The theoretical predictions for the inertial masses of the pion and muon are close to the experimental values. The parameters of the proposed geometrical configurations for the various particles are calculated and sketches of the proposed structures are shown in Figure 8.1. A theoretical value for the electron magnetic moment is obtained which is in agreement with measurement.

Chapter 8 reference

Particle Data Group 2014 *Review of particle physics*, Chinese Physics C Vol 38, No. 9 (2014) 090001

Chapter 9

Proton and neutron structures

9.1 Introduction

The proton model developed in Chapter 6 leads to predictions of the electronic charge and the gravitational constant in Chapter 8 which are close to the measured values. The equivalent spherical radius calculated from the volume for the proton is larger than the measured value. However a major drawback with the tapered proton model is that it assumes that there is no electric field leakage through the upper and lower surfaces, see Figure 9.1 (repeated from Chapter 6), and this clearly is not the case for the isolated particle.

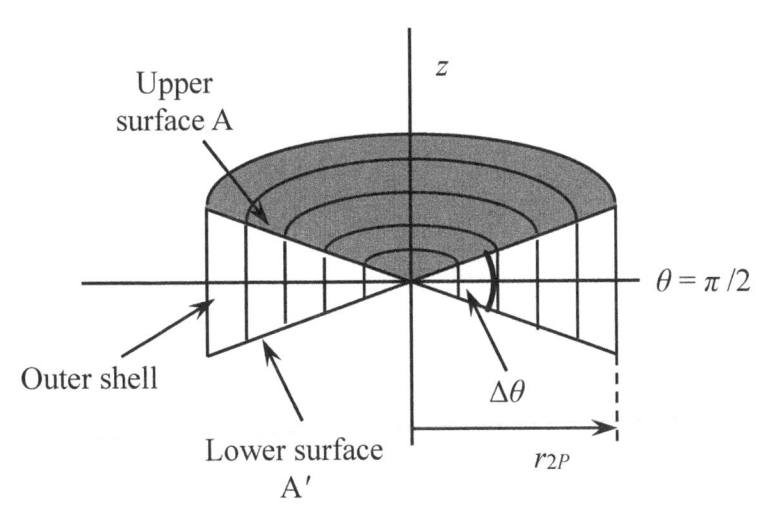

Figure 9.1 The original tapered proton model

It might be that the tapered structure is the one inside nuclei in which nucleons are tightly packed. For the isolated particle a key step is to introduce structure variants in which the leakage is reduced and which ensure better approximations to the spherical radial electric and gravitational field patterns required for applicability of the spherical shell model.

9.2 Structure of the proton

The features introduced are (a) to reduce the number of shells to a minimum in order to reduce the upper and lower surface areas and (b) to include a shell of negative charge split into upper and lower components to ensure a radial electric field near to the upper and lower outer surfaces and therefore allow radial fields throughout. A two shell structure is found to apply to the kaon, see Chapter 14. Here we examine the three shell structure for the proton shown in Figure 9.2.

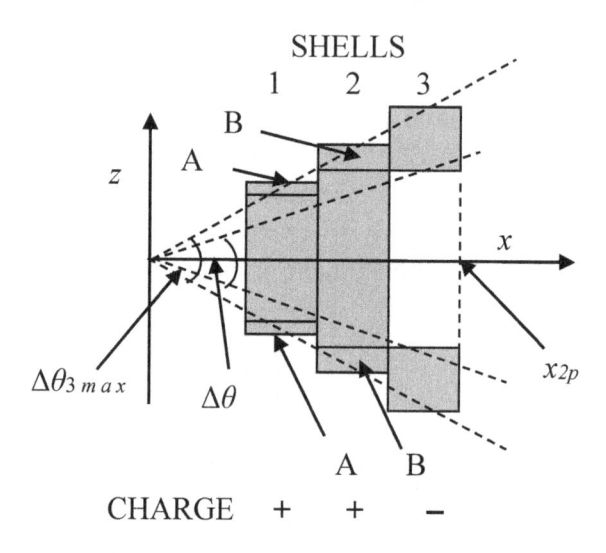

Figure 9.2 Three shell proton model

Formally we can take particular structures for the proton and, as in Chapters 6 and 8, predict values for the electronic charge and the gravitational constant. This is done below, with varying degrees of alignment between prediction and measurement. However Chapter 8 has already established that the new theory has the capability of predicting values for these constants. Our goal in this chapter is distinct from this. We require sufficiently detailed descriptions of the proton and neutron in order to predict the interaction potentials observed in scattering experiments and to estimate the binding energies of the low mass nuclei. Accordingly we assume values of the magnetic moments of the proton and the neutron in order to refine the selection of structures. The resulting values of the charge fractions in the proton

9.2 Structure of the proton

shells are close to the charge fractions observed for the quarks in the proton. We therefore identify the proton shells as the proton quarks and finalise the choice of proton structure on the basis of closeness to the measured charge fractions.

We consider in detail the configuration shown in Figure 9.2. It consists of positively charged shells 1 and 2, with charges q_1 and q_2 respectively. Shell 3 is split into upper and lower sections and is negatively charged, with charge q_3. The electric field lines emanating from the regions AA and BB terminate on the negative charge in shell 3. An approximation in the analysis that follows is that the charge contained in shells 1 and 2 net of the charge contained in AA and BB, that is

$$q = q_1 + q_2 + q_3 \tag{9.1}$$

gives rise to spherical radial electric field lines and they all pass through the outer surface of revolution subtending angle $\Delta\theta$ at the centre. This means that we can apply the field distributions of the spherical particle model to the Figure 9.2 structure. Thus the steady state electric potential varies as $1/r$ in spherical co-ordinates within each shell. However the steady state charge density at the centre of the ith shell (radius $r_{0i} = x_{0i} c/\omega_{0P}$) is such that, see Section 6.3,

$$\rho_{Bi} = \frac{x_{0j}^2}{x_{0i}^2} \rho_{Bj} \tag{9.2}$$

where ρ_{Bj} and x_{0j} are the values at the centre of the jth shell. We define the charge fractions by $c_i = |q_i|/q$ (which are all positive) so that from equation (9.1)

$$c_1 + c_2 - c_3 = 1 \tag{9.3}$$

We define a' by

$$c_1 + c_2 + c_3 = a' \tag{9.4}$$

Using (9.2) we have

Proton and neutron structures

$$q_1 = \rho_{B2}\pi^2 x_{02}^2 \Delta\theta_{3max}\left(\frac{c^3}{\omega_{0P}^3}\right) = q_2$$

where $\Delta x = \pi/2$ for filled shells and $\Delta\theta_{3max}$ is determined by equation (6.9), and so $c_1 = c_2$, and from (9.3) and (9.4) we have

$$c_1 = c_2 = (a' + 1)/4 \tag{9.5}$$

and

$$c_3 = (a' - 1)/2 \tag{9.6}$$

Now

$$q = |\rho_{B3}|\pi^2 x_{03}^2 (2\Delta\theta_{3max} - \Delta\theta_3)\left(\frac{c^3}{\omega_{0P}^3}\right)$$

where

$$\Delta\theta_3 = \Delta\theta_{3max} - \Delta\theta$$

and so

$$a' = \frac{2\Delta\theta_{3max} + \Delta\theta_3}{2\Delta\theta_{3max} - \Delta\theta_3}$$

and

$$\Delta\theta = \Delta\theta_{3max}\left(\frac{3 - a'}{a' + 1}\right) \tag{9.7}$$

The magnetic moment of a circular loop of wire with current i is $\mu_0 iS$ where S is the area contained by the wire. Hence the magnetic moment of the ith shell is given by $\mu_0 v_i |q_i| cr_{coi}/2$ where $v_i = \pm 1$ and is the product of the direction of circulation of charge (± 1) and the sign of the charge (± 1). We refer to the v_i as the particle circulation parameters. For a split shell the currents may be in opposing directions in upper and lower halves, resulting in $v_i = 0$. This is not used with the proton, but it is used with the neutron. When expressed in units of $\mu_0 e\hbar/2M_{0P}$, the nuclear magneton, the proton magnetic moment is

$$\mu_P = v_1 x_{01} c_1 + v_2 x_{02} c_2 + v_3 x_{03} c_3 \tag{9.8}$$

9.2 Structure of the proton

Equations (9.5) to (9.8) show a connection between μ_p and the c_i, a' and the various angles. Thus given μ_p, the x_{0i}, the v_i and $\Delta\theta_{3max}$, we have

$$a' = \frac{-v_1 x_{01} - v_2 x_{02} + 2v_3 x_{03} + 4\mu_p}{v_1 x_{01} + v_2 x_{02} + 2v_3 x_{03}}$$

and the c_i, $\Delta\theta$ and $\Delta\theta_3$ follow.

We consider three possible structures with differing values for x_{01}, x_{02} and x_{03}, Sets 1, 2 and 3 shown in Table 9.1.

Table 9.1 Values of the charge fractions and the angle $\Delta\theta_{3max}$ when the proton magnetic moment is 2.792847 nuclear magnetons for Sets 1, 2 and 3

	x_{01}	x_{02}	x_{03}	$\Delta\theta_{3max}$ rad	$c_1 = c_2$	c_3
Set 1	4.71	6.28	7.85	0.89	0.75	0.51
Set 2	6.28	7.85	9.42	0.82	0.71	0.41
Set 3	7.85	9.42	11.00	0.76	0.68	0.35

The possible values of x_{0i} are selected from those appropriate for the proton in Chapter 6 and the values of $\Delta\theta_{3max}$ are determined initially by equation (6.9). For example, for Set 3, the initial value of $\Delta\theta_{3max}$ is 0.76 radians. It is required that c_3 and $\Delta\theta$ are not to become negative and this restricts the options for the choice of the v_i. It is shown in the Book, Section 9.2, that $v_1 = 1$, $v_2 = -1$ and $v_3 = 1$ are suitable values. The resulting values for the charge fractions as functions of the magnetic moment are shown in Figure 9.3. The measured value of the proton magnetic moment is 2.792847 nuclear magnetons and for this value the predicted values for c_1, c_2 and c_3 are given in Table 9.1. We select Set 3 for the further work on the basis that it provides the closest agreement with the values $c_1 = c_2 = 2/3$ and $c_3 = 1/3$ required in the standard model and which are consistent with experiment.

Proton and neutron structures

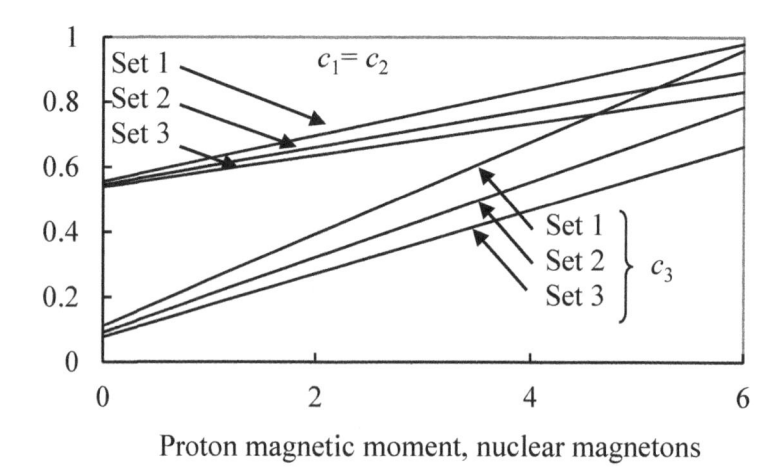

Figure 9.3 Theoretical values of the proton charge fractions versus the proton magnetic moment for Sets 1, 2 and 3 with $v_1=1$, $v_2=-1$, $v_3=1$ and $R=1$

There is further evidence justifying the use of Set 3 in Chapter 15. In Chapter 8 various results from the previous chapters are brought together in order to calculate theoretical values of some of the fundamental constants and other parameters, including the electronic charge and the gravitational constant. Because we have introduced a range of new proton models the analysis leading to the prediction of the electronic charge and the gravitational constant needs to be amended. Also, because of the near agreement of the Set 3 charge fractions and the standard model values, we wish to ensure that the proton magnetic moment has the experimental value. This means that we have to treat μ_P as an independent variable and this requires that it replaces one of the five constants amongst c, $\hbar$, ε_0, ω_{0P} and ω_{0e}. We choose ω_{0e} to be replaced. ω_{0P}/ω_{0e} and e are calculated as functions of μ_P using expressions in the Book, Appendix 9.A and the results are plotted in Figures 9.4 and 9.5.

9.2 Structure of the proton

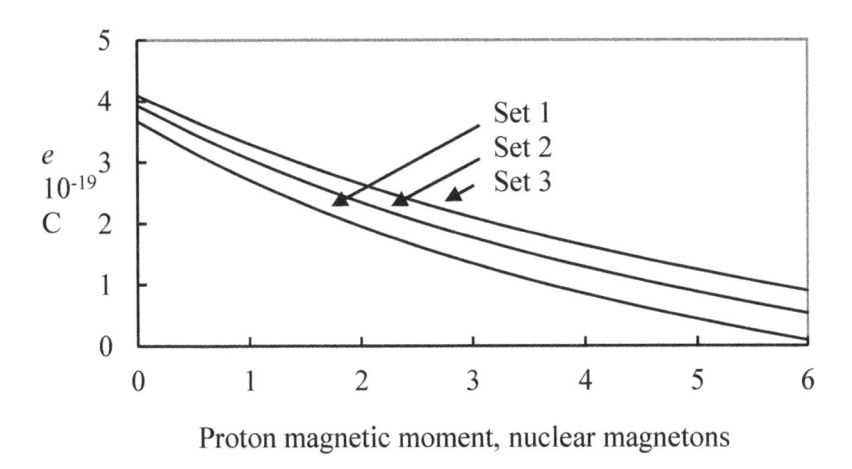

Figure 9.4 Theoretical values of electronic charge e versus the proton magnetic moment

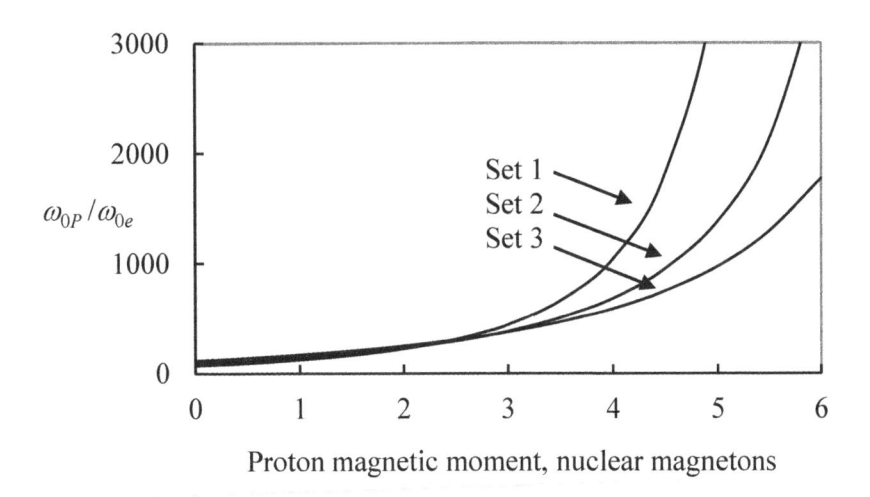

Figure 9.5 Theoretical values of ω_{0P}/ω_{0e} versus the proton magnetic moment

At $\mu_P = 2.792847$ for Set 3, the predicted ω_{0P}/ω_{0e} is well below 1836 and the predicted value for the gravitational constant is a factor of two below the measured value, see Table 9.2 with $\Delta_{3max} = 0.76$.

Proton and neutron structures

Table 9.2 The Set 3 structure predicted values of the proton charge fractions, the ratio ω_{0P}/ω_{0e}, the electronic charge and the gravitational constant using the measured value of the proton magnetic moment of 2.792847 nuclear magnetons

R	$\Delta\theta_{3max}$	$\Delta\theta$	$c_1 = c_2$	c_3	ω_{0P}/ω_{0e}	$\dfrac{e}{10^{-19}}$ C	$\dfrac{G}{10^{-11}}$ m³kg⁻¹s⁻²
1	0.76	0.36	0.68	0.35	345	2.21	3.5
2.3	1.52	0.72	0.68	0.35	1836	1.4	7.0

The original proton model of Chapter 6 used in Chapter 8 leads to a better alignment of the theoretical and experimental values for the gravitational constant. The discrepancies between the predicted and measured values of ω_{0P}/ω_{0e} and the gravitational constant can be resolved by adjustment of the proton model as follows. If $\Delta\theta_{3max}$ is increased by a factor of two, the theoretical value of G becomes close to the measured value. Doubling the angles and exceeding the condition of equation (6.9) is justified in Appendix D in which a solution is found for the proton oscillatory Field Equations in this case. The next change requires the introduction of the factor R defined by

$$m_B = R \frac{q}{2\pi\varepsilon_0 r_{2P} \Delta\theta}$$

(9.9)

This factor alters the value of the electric potential at the outer edge of the proton and is associated with an altered potential distribution in the vicinity of the proton. The combination of $R = 2.3$ and $\Delta\theta_{3max}$ being doubled changes the predicted value for e from 2.21×10^{-19} C to around 1.4×10^{-19} C and lifts ω_{0P}/ω_{0e} to 1836 – this data is collected together in Table 9.2, and further details are in Appendix D. Note that the charge fractions are not affected by these adjustments.

The conclusion from this section is that Set 3 with $v_1 = 1$, $v_2 = -1$, $v_3 = 1$ is preferred in which the structure is increased in size by

9.3 Structure of the neutron

doubling $\Delta\theta_{3max}$ to 1.52 radians and introducing $R = 2.3$. Another way of expressing this conclusion is that we have found a structure for the proton which gives rise to the measured value of the magnetic moment, the inertial mass of the electron, the charge of the electron, the gravitational constant, and the charge fractions of the quarks required by the standard model. Nevertheless R is an arbitrary factor and a more sophisticated analysis of the external potential distribution is required to establish the range of values of R which are realistic.

Thus we have established that an extended but partial spherical structure is possible for the proton. However it is more convenient in some cases to treat it as an extended cylindrical structure. There is discussion in Appendix D on where best to use partial spherical or alternatively cylindrical structures. An extended cylindrical structure is shown in Figure 9.6 in which the distance along z has been doubled compared to the extent limited by equation (6.9). For example the distance over shell 3 was previously 1.76 fm and is now doubled on Figure 9.6 to 3.52 fm. This structure is used in the next chapter on the overlap potential and the interaction of nucleons.

9.3 Structure of the neutron

A three shell configuration is examined in the Book, Chapter 9, for the neutron, based on the three shells of the configuration chosen for the proton, i.e. Set 3 with $x_{01} = 7.85$, $x_{02} = 9.42, x_{03} = 11.00$. As with the proton, segments of a spherical distribution of radial electric fields are assumed in order to use the results from the spherical particle model. The expression for the magnetic moment in equation (9.8) applies to the neutron. A number of alternative structures are examined in Chapter 9 of the Book. Of these the one with $v_1 = 0$, $v_2 = -1$ and $v_3 = 1$ is selected because of the proximity of the predicted magnetic moment of -2.48 nuclear magnetons to the measured magnetic moment value of -1.9 nuclear magnetons, and because it provides $c_1 = 0.32$, $c_2 = 0.63$ and $c_3 = 0.32$ values near to the standard model quark |charge| fractions 1/3, 2/3 and 1/3.

Proton and neutron structures

SHELLS
1 2 3

θ_{3max}

z

x

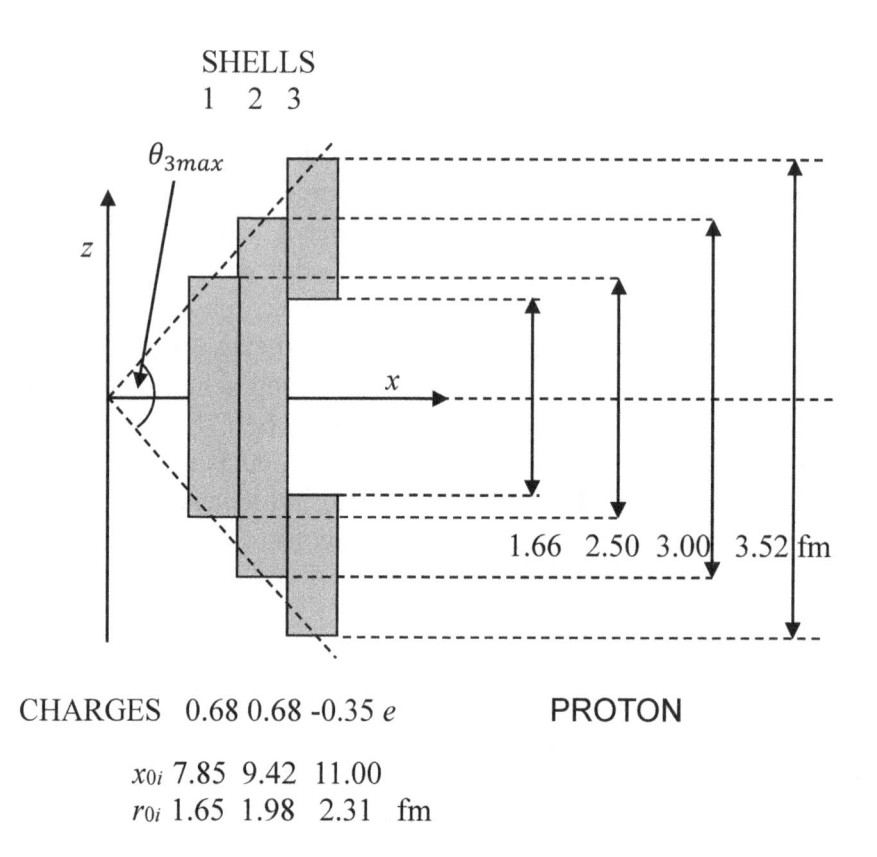

1.66 2.50 3.00 3.52 fm

CHARGES 0.68 0.68 -0.35 e PROTON

x_{0i} 7.85 9.42 11.00
r_{0i} 1.65 1.98 2.31 fm

Figure 9.6 Proposed proton structure. The particle volume is generated by rotating the cross section shown around the z axis. The particle consists of three shells with the charge fractions shown. All dimensions are in fm (10^{-15} m). Not to scale. The shell thickness is exaggerated

The cylindrical version of the structure is shown in Figure 9.7 and like the proton structure in Figure 9.6 the extent along z for each shell has been doubled compared to the structure limited by equation (6.9).

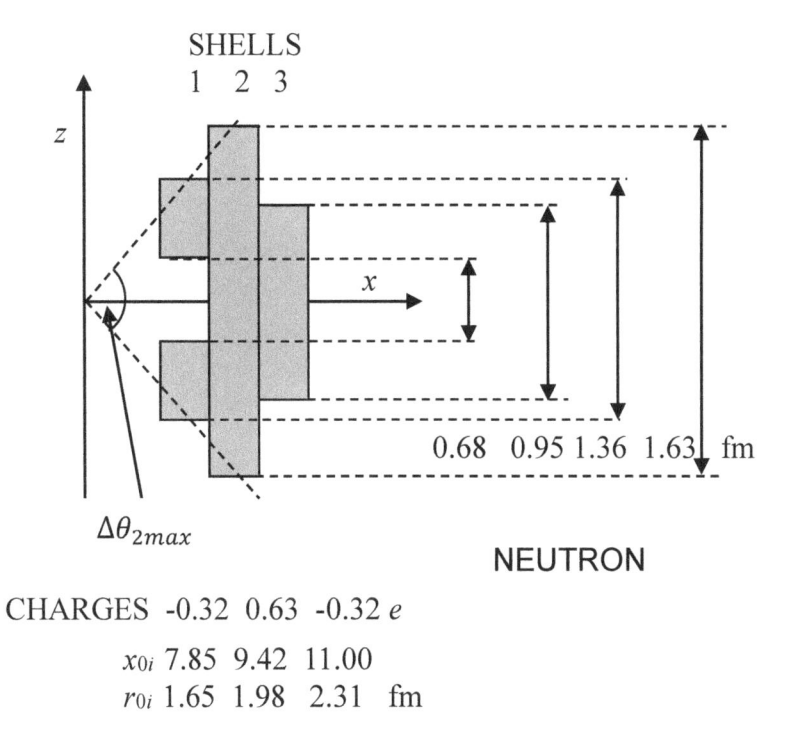

SHELLS
1 2 3

0.68 0.95 1.36 1.63 fm

$\Delta\theta_{2max}$

NEUTRON

CHARGES   -0.32 0.63 -0.32 e

x_{0i} 7.85 9.42 11.00

r_{0i} 1.65 1.98 2.31   fm

Figure 9.7 Proposed neutron structure. The particle volume is generated by rotating the cross section shown around the z axis. The particle consists of three shells with the charge fractions shown. All dimensions are in fm (10^{-15} m). Not to scale. The shell thickness is exaggerated

9.4 Summary

Structures containing three shells are developed in this chapter for the proton and neutron. It is proposed that the shells are in correspondence to the quarks of the standard model, and that the structures can be refined by using the nucleon quark charge fraction values and the measured values of the magnetic moments. Details of the structures are shown in Figures 9.6 and 9.7.

Chapter 10

The nuclear force

10.1 Introduction

The interaction of nucleons and their binding has attracted considerable interest ever since the discovery of the neutron and elucidation of the content of nuclei. A key aspect is the nature of the interaction potential energies involved in the scattering of one nucleon by another.

This chapter introduces a proposal for the origin of the force between nucleons. In the new theory it is proposed that there is a potential energy which arises as a consequence of the overlap of each shell in one nucleon with the corresponding shell in the other nucleon. The sign of this potential energy (strictly its gradient) determines whether the potential is attractive or repulsive. The degree of overlap determines the level of the potential energy involved and as a result this leads to predictions of the nucleon-nucleon interaction potential energies, to be compared with measurement data. The approach in this chapter uses the cylindrical structures determined for the proton and the neutron in the previous chapter.

10.2 The overlap model

The structures for the proton and the neutron arising from sections 9.2 and 9.3 are shown on Figures 9.6 and 9.7. We propose that interactions between nucleons can occur by the particles overlapping. In so doing we shall ignore any electrostatic or magnetic interactions due to the overlap of external fields with internal fields. We consider shell i in each particle. The overlap is indicated in Figure 10.1, illustrated by the overlap of a proton and a neutron. $\mathcal{V}_{Pi0}$ and $\mathcal{V}_{ni0}$ are the shell volumes prior to overlap. $\mathcal{V}_{Pi}$ and $\mathcal{V}_{ni}$ are the remaining non-overlapped volumes during overlap.

The nuclear force

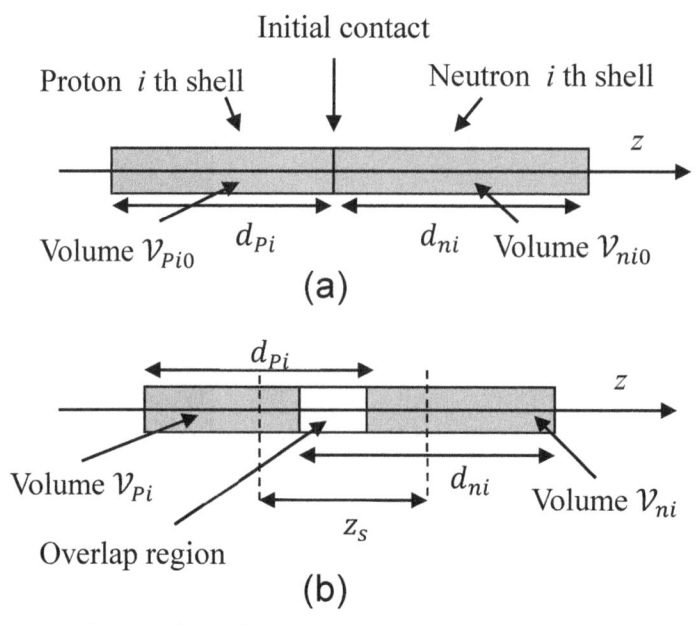

Figure 10.1 The overlap of proton and neutron shells. The z axis, for the purpose of illustration, is shown passing along the centre of the shell cross-sections. (a) Initial contact. (b) With an overlap region

It is shown below that potentials arise dependent on the overlap parameters

$$\beta_{ni} = (\mathcal{V}_{ni0} - \mathcal{V}_{ni})/\mathcal{V}_{ni0}$$
$$\beta_{Pi} = (\mathcal{V}_{Pi0} - \mathcal{V}_{Pi})/\mathcal{V}_{Pi0}$$

indicated on Figure 10.1. z_s is the separation between shell centres and β_{ni} and β_{Pi} are functions of z_s.

Consider the quasi-steady state charge and gravitational mass densities in the overlap region. They are (see section 3.4) of the form

$$A_\rho \rho_{A1} \rho_{A1}^* + A_\rho \rho_{A2} \rho_{A2}^* + A_\rho \rho_{A1} \rho_{A2}^* \cos \Delta \omega t$$
$$A_m m_{A1} m_{A1}^* + A_m m_{A2} m_{A2}^* + A_m m_{A1} m_{A2}^* \cos \Delta \omega t$$

where $\rho_{A1}, \rho_{A2}, m_{A1}, m_{A2}$ refer to particles 1 and 2 respectively, $\Delta \omega$ is the frequency difference between the shells and the angular frequency sum terms have been dropped. The difference terms must correspond to standing wave solutions of the oscillatory Field Equations. If for some $\Delta \omega$ a mode can be supported, then as $\Delta \omega$ is reduced, a point will

10.2 The overlap model

be reached where the overlap region (or some more extended region) has insufficient extent to continue supporting the mode. If such modes cannot be supported then the overlapping particles must be coherent, and the resulting steady state density levels in the overlap region depend on the interference oscillatory waveform amplitudes. We examine the coherent situation and this is key to there being an overlap potential.

It is important to determine whether the interference is to be constructive or destructive. If the overlapping shells have their currents in the same direction, then constructive interference is required, and if they are in opposing directions destructive interference is required. For identical nucleons the currents are in the same direction when spins are parallel, and in opposing directions when antiparallel. Thus for the spin-up/spin-up situation, the circulation parameters are the same for each shell and there is constructive interference between shells 1, between shells 2 and between shells 3. For spin-up/spin-down there is destructive interference.

We examine these considerations in more detail and we consider the overlap of a proton and neutron. We need to be clear how we define the circulation parameters. We use the direction of the spin-up axis to define the direction of z. The circulation parameters are now defined with respect to z. We illustrate these conventions in Table 10.1 for shells 2.

Table 10.1 The type of interference (constructive c or destructive d) resulting from the overlap of shells 2 of a proton and a neutron

Proton	v_{P2}	Neutron	v_{n2}	c or d
Spin up	-1	Spin up	-1	c
Spin up	-1	Spin down	1	d
Spin down	1	Spin up	-1	d
Spin down	1	Spin down	1	c

When the proton and neutron are both spin-up, their circulation parameters are those introduced in the previous chapter. Since $v_{P2} = -1$ and $v_{n2} = -1$, then constructive interference is required. When the neutron is spin down the circulation parameter becomes 1. So it is

The nuclear force

the circulation parameters which determine whether constructive or destructive interference takes place as shown in Table 10.1.

If we put $z \rightarrow -z$ in the lower two entries, they can be described by the upper two entries and the table can be reduced to these two cases. Using the conventional quantum mechanical notation for spin states, the first of these corresponds to the case described by $s = 1$ with $m_s = 1$ and the second by $s = 0$ with $m_s = 0$. We can now proceed to consider the interaction of a proton with a neutron, a proton with a proton and a neutron with a neutron. In each case shell 1 of one particle overlaps the shell 1 of the other, shell 2 with shell 2 and shell 3 with shell 3. The detail leading to the determination of whether there is required to be constructive or destructive interference is shown in Tables E.1, E.2 and E.3 in Appendix E.

We consider next how the required interference can occur. We need to examine how the oscillatory waveforms can adjust to provide the required constructive or destructive interference. We can write the oscillatory charge density amplitude using cylindrical co-ordinates for one particle in the form (see Chapter 2),

$$\rho_A = R(r_c)Z(z)\exp(-i\phi/2)$$

for $0 < \phi < 2\pi$ and corresponding to 'spin-up'. We consider the cases where the phase differences between shells are described by differences in their spin function arguments. So overlapping spin-up/spin-up particles can interfere constructively. Similarly so can spin-down/spin-down particles where the spin function becomes $\exp(i\phi/2)$. However there are other cases to consider. Examination of Table E.3 in Appendix E shows that destructive interference can be required as part of the spin-up/spin-up overlap. Also with overlapping spin-up/spin-down particles one particle has the spin function $\exp(-i\phi/2)$ and the other has $\exp(i\phi/2)$ and constructive or destructive interference may be required.

Thus we need to examine the four cases: $s = 1$, $m_s = 1$ with constructive or destructive interference, $s = 0$, $m_s = 0$ with constructive or destructive interference. It is shown in Chapter 10, Section 10.2 of the Book how the spin functions can be chosen to satisfy these requirements, consistent with the system spin state, i.e. the spin function choices in the Book maintain s and m_s at the required values throughout the overlap of a pair of particles.

10.3 The overlap potential energies

10.3 The overlap potential energies

We can now examine how the potentials are introduced as a result of particle overlap. Central to the analysis is the overlap parameter, which is specific to the shell combination which is interacting. Examples of overlap parameters as functions of z_s are shown in Appendix E on Figures E.1, E.4 and E.7. We illustrate the principles for obtaining expressions for the potential energy by examining the overlap of two identical positively charged proton shells, where various parameters are shown on Figures 10.2 and 10.3.

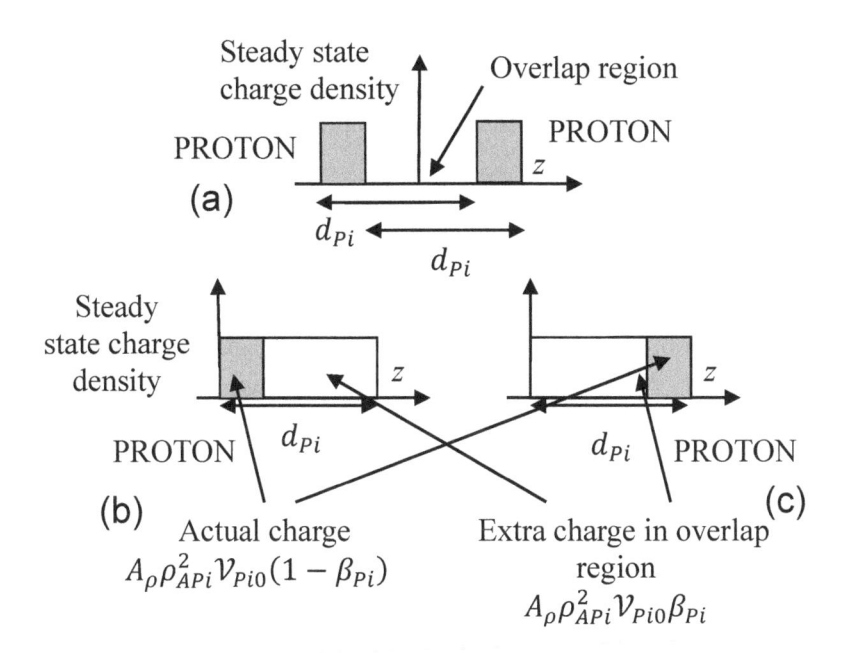

Figure 10.2 Destructive interference with two protons. (a) Two reduced volume particles separated by the overlap region. (b) and (c) Representations of the reduced volume particles. The extra charge in the overlap region is cancelled by the effect of the negative gravitational potential

The nuclear force

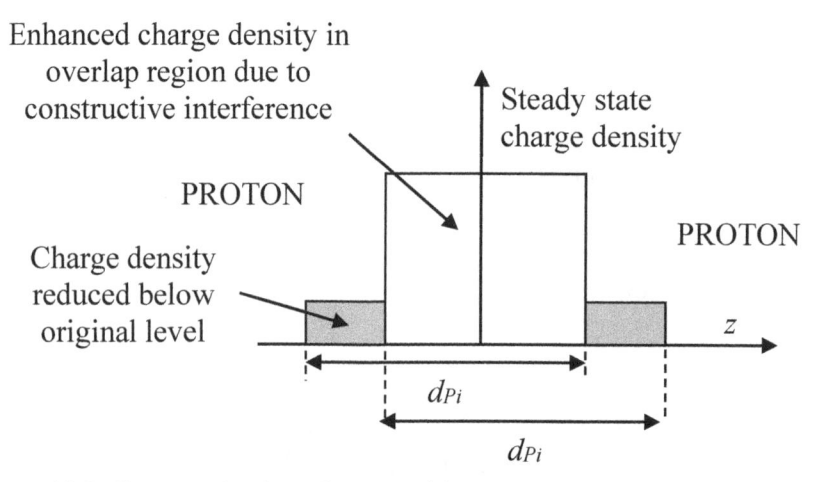

Figure 10.3 Constructive interference with two protons

The detail for all cases is in Appendix 10A in the Book. We consider shell i in each particle and we consider first destructive interference and then constructive interference. We turn our attention to Figure 10 (a). There is destructive interference in the overlap region; zero oscillatory amplitudes mean zero steady state levels. The effective volume of each shell reduces as the overlap increases and the steady state charge density in each shell increases, indicated by the shaded regions. The charge of each shell, which remains constant, is given by

$$q_{Pi} = \int_{v_{Pi}} \rho_{Bi0} \left(\frac{1}{1 - \beta_{Pi}}\right) dv$$

where

$$\rho_{BPi0} = \frac{q_{Pi}}{v_{Pi\,0}}$$

is the original level of the steady state charge density. A reduced volume particle can be represented as an original volume particle with the new value of steady state charge density cancelled in the overlap regions by a negative gravitational potential, see Figure 10.2 (b) and (c), and so

10.3 The overlap potential energies

$$q_{Pi} = \int_{V_{Pio}} \left\{ \frac{q_{Pi}}{V_{Pio}} \left(\frac{1}{1-\beta_{Pi}} \right) + U \right\} dV$$

and so

$$U = -\frac{q_{Pi}\beta_{Pi}}{(1-\beta_{Pi})V_{Pio}}$$

Similarly an electric potential V can be introduced to account for the reduction in the volume occupied by the steady state gravitational mass given by

$$V = -\frac{M_{Pio}\beta_{Pi}}{(1-\beta_{Pi})V_{Pio}}$$

where M_{Pio} is the gravitational mass in the ith shell. Thus the total potential energy, when $\beta_{Pi} \ll 1$ (to ensure that the potentials are small compared to the particle densities) and taking into account a factor ½ in forming the joint system, is

$$-\left(\frac{2M_{Pio}\, q_{Pi}}{V_{Pio}} \right) \frac{\beta_{Pi}}{1-\beta_{Pi}}$$

This describes an attractive interaction when q_{Pi} is positive and a repulsive interaction when q_{Pi} is negative.

When there is constructive interference there must be a reduction in the steady state charge density in the non-overlapped regions, now ρ_{BPi}, to balance the local increase in the density in the overlap volume, see Figure 10.3. The charge in the overlap region is $4\rho_{BPi}\beta_{Pi}V_{Pio}$ split between the two particles. The charge contribution to each particle is further split between that due to ρ_{BPi} in the overlap region and that due to the gravitational potential contained entirely within in the overlap region. If we now treat U as a field existing throughout the particle, then $U = \rho_{BPi}\beta_{Pi}$ and

$$q_{Pi} = \int_{V_{Pio}} \{\rho_{BPi} + U\} dV = \rho_{BPi} V_{Pio}(1+\beta_{Pi})$$

Hence

The nuclear force

$$\rho_{BPi} = \frac{\rho_{BPi0}}{1 + \beta_{Pi}}$$

$$U = \frac{\rho_{BPi0}\beta_{Pi}}{1 + \beta_{Pi}}$$

Similarly

$$m_{BPi} = \frac{m_{BPi0}}{1 + \beta_{Pi}}$$

$$V = \frac{m_{BPi0}\beta_{Pi}}{1 + \beta_{Pi}}$$

Taking into account the factor of ½, the total PE is

$$\frac{2M_{Pi0}\, q_{Pi}\beta_{Pi}}{\mathcal{V}_{Pi0}(1 + \beta_{Pi})}$$

This describes a repulsive interaction when q_{Pi} is positive and an attractive interaction when q_{Pi} is negative.

Appendix 10A in the Book develops the analysis by starting with the general p-n cases, dealing with positively and negatively charged shells and relaxing the condition that the potentials are small. Tables E.1, E.2 and E.3 summarize various parameters of the p-n, p-p and n-n interactions. The tables refer to the various formulae in the Book which have been used to calculate the potential energies, and to the figures in Appendix E displaying the results.

The major features of experimental data on low energy nucleon – nucleon scattering have been captured in phenomenological potential energy functions versus radius in spherical co-ordinates. Examples of such are shown in Figure 10.4.

Curves A and B show the lower and upper extent in radius of these functions when the potential energy is negative and C shows the typical observed results when the potential energy is positive (see Krane (1988, p107)). Figure 10.4 also shows the total potential energy for p – n, p – p and n - n interactions taken from Figures E.2, E.3, E.5, E.6, E.8 and E.9 for comparison with the experimental results.

10.3 The overlap potential energies

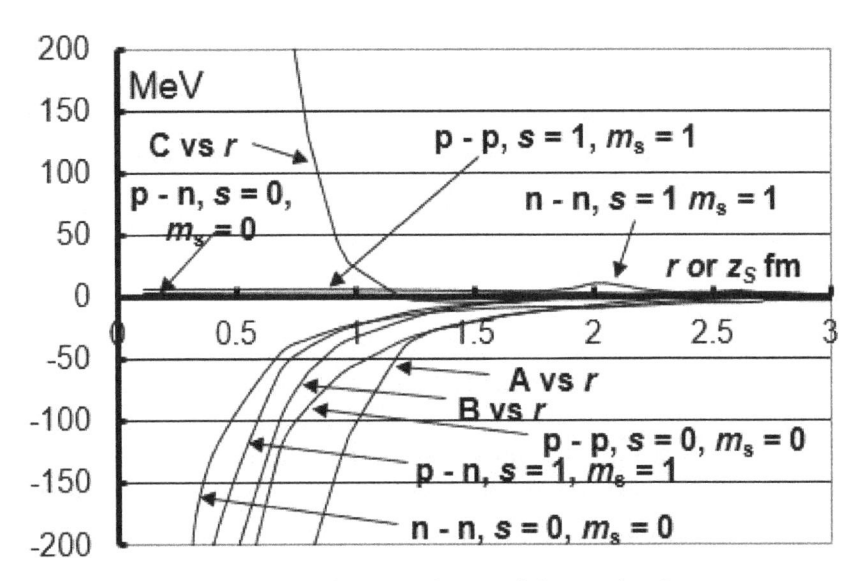

Figure 10.4 Comparison of the total potential energies for p − n, p − p and n − n interactions with the observed values ranging between A and B, together with observed C. $R = 2.3$ and $e = 1.4 \times 10^{-19} C$

All these theoretical results have used $e = 1.4 \times 10^{-19} C$ and $R = 2.3$. Note that the comparison is between predicted potential energy versus z and experimental potential energy versus r. A more comprehensive theoretical account needs to extend the overlap analysis to the full 4π solid angle for comparison with the experimental data.

From these comparisons we draw the following conclusions.
(1) For negative potential energies, the doubled z_s theoretical results show similar variations with distance to the experimental results.
(2) At large z_s in the experimental results there is low energy structure in the potential energy functions. There is similar detail in the theoretical results. Maybe actual nucleon structures can be derived from the experimental data.
(3) The factors of the form $q_{Pi} M_{Pi0}/V_{Pi0}$ appear to predict the right sort of magnitudes for agreement with experiment.
(4) The theoretical variation of potential energy with z_s arises from the effect of overlap and the approximate agreement with experiment gives credence to the overlap mechanism.
(5) The experimental evidence shows that there is an intricate

The nuclear force

dependence on spin states, and this is also the case with the theoretical analysis.

(6) The theoretical results predict positive potential energies in some cases but they do not show a high magnitude positive potential energy as exhibited by the experimental curve C. However C is associated with orbital angular momentum $l = 1$ and our analysis, being restricted to motion along the z axis, does not include this case.

Further work will need to involve:

(1) Extending to three dimensions the analysis of the nucleon-nucleon overlap in order to obtain interaction potentials in three dimensions.
(2) Refining (or radically altering) proton and neutron structures to align experimental and theoretical scattering results, and then to examine whether these structures emerge from a revised theoretical analysis.
(3) An initial attempt is made to model structures for the light nuclei based on overlap potential energies in Chapter 11 of the Book, and this needs to be extended or superseded.

10.4 Summary

It is proposed that interactions between nucleons can occur by the particles overlapping. This can give rise to interference between the particles' oscillatory waveforms. The interference can be either constructive or destructive. The overlapped system can be represented by particles filling their original volumes together with electric and gravitational potentials. The predicted potential energies are in approximate agreement with the experimental results.

Chapter 10 reference

Krane K S 1988 *Introductory nuclear physics* (Wiley: New York)

Chapter 11

Anti-particles

11.1 Introduction

Do anti-particles have positive or negative gravitational mass? The purpose of this chapter is to examine some of the consequences of our models for anti-particles. In the Book, Chapter 2 Section 2.4, we have commented that new particles can be described by taking a given particle solution and putting $A_m = -A_m$, $A_\rho = -A_\rho$. The oscillatory solutions remain the same, but the particle steady state electric charge and gravitational mass densities change sign. We have dubbed these anti-particles. The major feature is that these particles have a negative gravitational mass. This leads to a number of issues which we discuss below. In Chapter 6 Section 6.5 of the Book we also imply the existence of an anti-particle with a change in sign of the electric charge but retaining a positive gravitational mass. We examine this alternative in detail below and conclude that, although both versions of anti-particle are possible, it is the latter one, the positive gravitational mass one, which prevails in our universe.

11.2 Anti-particles with negative gravitational mass

It is pointed out in Chapter 4 that the expression for $\hbar$ is the same whether A_m is positive or negative since it is multiplied by M_0 in the expression. This means that since the inertial rest mass is defined by $\hbar\omega/c^2$ then the inertial mass is positive when the gravitational mass is negative. There are a number of consequences which result from the adoption of the negative gravitational mass model for anti-particles:

(1) They are repelled by the positive gravitational masses in our universe.

(2) When a particle and anti-particle annihilate, the particles disappear leaving zero net gravitational mass. If the energy is taken away by photons, one photon will need to have positive

Anti-particles

gravitational mass and the other negative gravitational mass.

(3) When a particle absorbs energy from a photon, gravitational mass is transferred from the photon to the particle. Negative gravitational mass will need to be transferred to an anti-particle when energy is transferred to the anti-particle. If there are no negative gravitational mass photons available, this cannot happen.

(4) Remembering that the issue here is whether anti-particles with negative gravitational mass are prevalent in our universe, and since such anti-particles are predicted from our theory, then regardless of whether we conclude that this prevalence is or is not the case, the speculation remains as to whether there are universes within the multiverse which are composed of particles with negative gravitational mass.

There are a number of questions and difficulties:

(1) The first concerns the creation of an e^-e^+ pair. If e^+ has positive gravitational mass, this can happen via an incoming gamma, (in the presence of a nucleus to satisfy momentum conservation). However if e^+ has negative gravitational mass, two gammas will be required, one with positive gravitational mass and the other negative gravitational mass. Momentum conservation can be satisfied without having a nearby nucleus. However this mechanism does not seem to happen experimentally, and so the evidence supports the conclusion that these particles both have positive gravitational mass.

(2) The Λ^0 has an inertial mass of around 1.116 MeV c^{-2} (Particle Data Group 2014) compared to 938 MeV for the proton. Consider the following decays:

$$\Lambda^0 \rightarrow p + \pi^-$$
$$\pi^- \rightarrow \mu^- + \bar{\nu}_\mu$$
$$\mu^- \rightarrow e^- + \bar{\nu}_e + \nu_\mu$$

Since π^- is an anti-particle with, here, an assumed negative gravitational mass, then the lambda is required to contain a net gravitational mass which is less than that of the proton despite having a higher inertial mass. The second and third decays imply that the muon and electron contain negative gravitational mass which is contrary to what we have predicted so far. Are

11.3 An alternative model for anti-particles with positive gravitational mass

there electron variants with negative gravitational mass generated from negative pion decay and what happens to them? These are major difficulties.

(3) Is space permeated with negative gravitational mass photons? Do negative gravitational mass photons interact with atoms e.g. do they occur in laser radiation, i.e. are there negative gravitational mass photon waveforms? There is $2A\Delta\omega$ gravitational mass to be accounted for when the atomic electron loses $A\Delta\omega$ and the photon waveform loses $A\Delta\omega$. Conceivably these could be absorbed by two zero point energy photon waveforms changing from negative gravitational mass to positive gravitational mass at constant energy. However this a complex process which would be expected to have different properties from normal atom - photon interactions.

(4) Negative gravitational mass photon waveforms will bend in the opposite direction from positive gravitational mass photon waveforms in a gravitational field, and observation to date suggest that such waveforms are not prevalent.

11.3 An alternative model for anti-particles with positive gravitational mass

A different type of anti-particle can be obtained by putting $A_\rho = -A_\rho$ but retaining a positive sign of A_m. The oscillatory solutions remain the same. The steady state solutions remain the same, based on the following. For baryons and some mesons it is sufficient to examine the solutions for the spherical particle model in Appendix A to Chapter 6. Since for both the steady state charge density and the gravitational mass density, Poisson's equation is approximated by Laplace's equation, then solution m_B can remain the same, and ρ_B can change sign. In detail there will be changes in the exact solution of Poisson's equation. Nevertheless, we conclude that there will be particle counterparts with change of sign of electric charge but a positive gravitational mass. This conclusion applies to the cylindrical shell models for baryons. Where meson models are based on cylindrical or spherical shells, then the same conclusion applies. Pions need further attention and this is dealt with in Chapter 14, Section 14.6.

So what about leptons? Maybe for the positron it is sufficient to

Anti-particles

increase the positive charge section of Figure 12.6 in the Book, retaining a positive gravitational mass and with a positive external steady state gravitational mass density. We can apply the same mechanism to obtain the positively charged muon.

There are consequences and issues which follow from adopting the positive gravitational mass anti-particle:

(1) The difficulties with the π^- decay discussed above in the previous section are removed.
(2) A negatively charged shell with positive gravitational mass already exists in the nucleon models dealt with in the Book (Chapter 9) and identified as the down quark (Chapter 12). Therefore this provides the exemplar for achieving negatively charged counterparts of positively charged shells. Further analysis is to be found in Appendix G, Section G.6, which explains why the angular frequency and hence the inertial mass and gravitational mass of a charged anti-particle is the same as that of the particle.
(3) Models are required which differentiate between neutrinos and anti-neutrinos (reversal of inner and outer layers, reversal of positive and negative charges, minute residual electric charge either positive or negative. Do they contain negative gravitational mass?)

11.4 Conclusions

The difficulties for the prevalence of negative gravitational mass anti-particles listed above are severe, and the difficulties for positive gravitational mass anti-particles are minor in comparison. So we conclude that we need to apply the case where anti-particles have positive gravitational mass in the interpretation of experimental data. Nevertheless the results from future crucial experiments and observations could undermine this decision. Examples are with the spectroscopy of anti-hydrogen and whether anti-particles rise or fall in the Earth's gravitational field. We also conclude that negative gravitational mass anti-particles could exist as dominant particles in other universes within the multiverse.

Chapter 11 reference

11.5 Summary

There are two possibilities in accounting for anti-particles in the new theory. One involves a negative gravitational mass, the other a positive gravitational mass. The conclusion is that anti-particles observed in experiments so far have positive gravitational mass.

Chapter 11 reference

Particle Data Group, 2014 *Review of particle physics*, Chinese Physics C Vol 38, No. 9 (2014) 090001

Chapter 12

The origin of probability in quantum mechanics

12.1 Introduction

In the Book the new theory is developed in the first four chapters to encompass quantum mechanics and the steps involved are summarised in this book in Chapters 1 to 4. The connection is made by showing that basic features of quantum mechanics follow from the new theory. In Chapter 13 of this book we extend the new theory to encompass quantum field theory (QFT) and the subset of that field, quantum electrodynamics (QED). However to place the step from Chapter 4 to Chapter 13 on a formal basis, we need to show that the postulates of quantum mechanics follow from the new theory. This is the subject of this chapter. It precipitates the issue as to whether or how probability is to be introduced into the new theory. We take the situation described in Chapter 4 of the Book with regard to non-relativistic quantum mechanics and put it on a more formal basis. Importantly we show how probability is introduced into quantum mechanics.

This chapter is structured as follows. First we briefly refer to the major issues of probability and hidden variables in quantum mechanics. We then discuss how probability is introduced, within our theory, into quantum mechanics, with reference to Rutherford scattering, spontaneous emission, and the double slit experiment. Next we show the extent to which the postulates of quantum mechanics follow from the position reached in Chapter 4 of the Book. In Chapter 4 of the Book we point out that there is no agreement on the exact statement of the postulates of quantum mechanics. The postulates set out by Dicke and Wittke (1960) are comprehensive and these are the ones that we use here. We conclude that, although there are differences in interpretation, the postulates follow from our theory.

12.3 Probability, scattering and other processes

12.2 Probability and hidden variables

There are a number of bizarre aspects of quantum theory which seem to defy rational explanation. Examples lie with the apparent collapse of the wave function when making measurements, the phenomenon of entanglement, and interference effects such as exhibited by quantum eraser experiments. Debate centres on how the theoretical results are to be given a physical interpretation. Extensive reviews and discussion are to be found in accounts by Al-Khalili (2012), Cox and Forshaw (2012), Baggott (2011), Kumar (2008) and Penrose (2005). The role of probability in quantum mechanics was the subject of the famous debate between Bohr and Einstein. The school of thought led by Bohr with their Copenhagen probabilistic interpretation of the Schrödinger wave function prevailed. It has been a topic of contention since, with theoretical points advanced and experiments conducted in order to resolve the principles involved.

A major aspect of the debate has concerned the existence of hidden variables. The sequence of events is set out by Kumar who identifies key contributions made by von Neumann, Einstein Podolsky and Rosen, Bohm, Bell, Aspect et al, Leggett, and Aspelmyer and Zeilinger. More recent work, based on theoretical work by Kochen and Specker (1967) and Klyachko et al (2008), has led to experiments by Lapkiewicz et al (2011) from which it is concluded that hidden variables cannot explain certain types of photon measurements. We show in this chapter how probability can be introduced, and in Appendix F how the new theory can address some of the contentious issues.

12.3 Probability, scattering and other processes

In Chapter 4 we introduce the wave function ψ_D. In the Book Chapter 4 we use the notation $\psi_D = \psi_T$ for a particle in translational motion and $\psi_D = \psi_W$ for a particle in a potential well. These are wave functions which describe actuality. The basic particle solution fills the particle's volume with electric charge and gravitational mass. These wave functions distribute the particle beyond its basic boundaries. So an electron in an atom is distributed over the atomic volume. A wave packet describes the distribution of a moving particle. We make measurements on these various entities using a process, where there is

The origin of probability in quantum mechanics

an input or inputs and an output or outputs. However there are many things that we do not know regarding the inputs, e.g. phase of a wave function, position in space of a trajectory, the parameters of photons which interact with a particle. These are just the known unknowns. So we choose to treat the situation statistically. In so doing this is how we introduce probability into quantum mechanics and we do so via a probabilistic factor ψ_P introduced into the wave function. It will be seen that Postulate 5 in Appendix F on probability applies to ψ_P but not to ψ_T and ψ_W. In order to develop this concept we examine three specific cases: Rutherford scattering, spontaneous emission, and the double slit experiment.

Rutherford scattering. QFT is concerned with input and output states and therefore is principally concerned with various forms of scattering. Here we restrict the examination to the non-relativistic scene. In Chapter 13 which deals with QED and QFT the examination is broadened to the relativistic scene.

First we consider the input to scattering. We start with the wave packet introduced in Chapter 4, for a charged particle, and we examine the particle scattered off say a stationary nucleus as in Rutherford scattering, in a direction defined by θ, ϕ in spherical co-ordinates with the nucleus at the origin. The wave function is $\exp(-ikz)$ when of infinite extent in z, but otherwise we will use the wave packet described by $\psi_T(z)$ such that $\int \psi_T(z) \psi_T^*(z) dz = 1$ over the packet. The steady state electric charge distribution is given by $\psi_T(z)\psi_T^*(z) * (\psi_R \psi_R^* * \rho_B)$ where $*$ denotes convolution, ψ_R is the radial wave function introduced in Chapter 4 in the Book, and the integral of the steady state charge density of the basic particle, ρ_B, over the particle volume is the particle charge. The offset of the particle's initial trajectory from the central axis going through the nucleus is the impact parameter so that in cylindrical co-ordinates r_c is the impact parameter.

We could have attempted a deterministic solution, taking into account a specific input, the various states of any interacting photons and so on. The Field Equations suggest that such a solution is possible and of course there is already a solution: it is the classical analysis leading to the Rutherford formula for the scattering cross-section.

We now *choose* to consider an ensemble of particles distributed over all impact parameters out to a beam radius r_B such that the probability per particle per unit area for the impact trajectory being

12.3 Probability, scattering and other processes

parallel to z at r_c, ϕ is $\psi_{P1}\psi_{P1}^*$ where in general ψ_{P1} is a function of r_c and ϕ. This step makes the quantum mechanical treatment stochastic, i.e. a form of statistical mechanics. In order for the particle to be somewhere in the beam the integral of $\psi_{P1}\psi_{P1}^*$ over the beam area is unity. We can write the charge density as $\psi_{P1}\psi_{P1}^*\psi_T(z)\psi_T^*(z) * (\psi_R\psi_R^* * \rho_B)$ and the integral of this over the beam area and z is the particle charge.

We now consider the output. Just as for the input where we have a wave packet but do not know its impact parameter, then the output will be a wave packet described by a translational wave packet, but since we have a stochastic input, we have no deterministic way of calculating the output direction. So we describe the output by the wave function $\psi_{P2}(\theta, \phi)$ where ψ_T is now along r. So the scattered particle has a probability distribution $\psi_{P2}(\theta, \phi)\psi_{P2}^*(\theta, \phi)$ per unit solid angle.

So how do we calculate $\psi_{P2}(\theta, \phi)$? If we distribute the input using ψ_{P1} then an actual input which includes all possible inputs will be obtained, and ψ_{P2} will describe an actual output which includes all possible outputs. Since these are actual distributions, then the time independent Schrodinger equation will apply. We continue exactly as in conventional quantum mechanics along the following lines, see Dicke and Wittke (1960 pp 291 – 296). We put the input in the form $A\exp(-ikz)$ and v is the scattered wave. The total wave function is the sum $v + A\exp(-ikz)$. v is eventually put in the form of an amplitude (leading to the probability versus θ, ϕ) multiplying a radial wave function. Using Schrödinger's equation and the Born approximation then

$$\nabla^2 v + k^2 v = \frac{2\mathcal{M}V}{\hbar^2}\exp(-ikz)$$

where in spherical co-ordinates,

$$\nabla^2 v = \frac{1}{r^2}\frac{\partial}{\partial r}\left(r^2\frac{\partial v}{\partial r}\right) + \frac{1}{r^2 \sin\theta}\frac{\partial}{\partial \theta}\left(\sin\theta\frac{\partial v}{\partial \theta}\right) + \frac{1}{r^2 \sin^2\theta}\frac{\partial^2 v}{\partial \phi^2}$$

where V is the scattering potential energy and $\mathcal{M}$ is the particle inertial mass. The mathematics involves the use of a Green's function in solving for v and this leads to an expression for the differential cross section which is proportional to $1/\sin^4(\theta/2)$. Hence the new theory leads to the Born approximation treatment for Rutherford scattering.

The origin of probability in quantum mechanics

The input has been chosen to be in stochastic form and the output has also been cast in stochastic form.

Rutherford scattering is treated as a standard piece of analysis in classical mechanics, see for example Goldstein (1959 pp81 - 85). By appropriate choice of the potential, the quantum mechanical expression for the differential cross section reproduces the classical expression (Dicke and Wittke 1960 p296). Goldstein (p82) discusses the connection between the impact parameter and the scattering angle in the classical mechanics treatment. However in our theory we have introduced probability. We do the quantum mechanical analysis and we obtain a probabilistic answer. We do not know the impact parameter; we cannot deduce within quantum mechanics the connection between the impact parameter and output direction. It is tempting to guess that there is a connection from a comparison of the quantum mechanics and classical mechanics treatments. However in this form of our theory and in conventional quantum mechanics there is no connection made between impact parameter and output direction.

There is an alternative quantum mechanics treatment of Rutherford scattering. We have the concept of $\psi_{P2}(\theta, \phi)$ which is stochastic. We can say that $\psi_{P2}(\theta, \phi)$ provides the coefficients for a set of output states. We can calculate ψ_{P2} using time dependent perturbation theory, leading to Fermi's Golden Rule. Das and Ferbel (1994) give an account (pp 25 – 26) where the process also leads to the Rutherford formula. Hence the time dependent form of Schrödinger's equation and the use of time dependent perturbation theory apply in this case. Given a generalisation of this, discussed in Appendix F, we can apply the time dependent form of Schrödinger's equation and time dependent perturbation theory.

Spontaneous emission. Given a particle in a potential well in an initial state described by ψ_W, there may be a number of lower energy states to which it may transfer. It requires the presence of a photon waveform corresponding to the appropriate energy difference for an allowed output state to result. Since it is not known in advance which photon waveform may be present, we choose to treat the problem statistically using conventional time dependent perturbation theory. As a result the expression for the transition rate to any one state is independent of the transition rates to other states.

This independence of the transition probability for any one output state is important in that it allows one to think in terms of allowed and

12.5 The postulates of quantum mechanics

forbidden output states and one can conceptually add or remove such states without affecting the transition rate or scattering cross section to some state under consideration, and to do calculations without knowing what the complete set of allowed output states is.

Double slit experiment. This is discussed in Appendix F, Section F6. The double slit experiment with its quantum mechanical interference pattern precipitates major issues underlying the physical interpretation of quantum mechanics (e.g. which slit did the particle travel through?). The interpretation in the new theory is dealt with in Appendix F.

12.4 Measurement

If we take a measurement to be a process, with input and output, such as scattering, we come to the following conclusions. The possible outputs from the process are contained in a set of actual states, each state described by a wave function which actually distributes the particle. ψ_T and ψ_W are examples of such wave functions. With Rutherford scattering the output particles travel in particular directions described by θ, ϕ. With spontaneous emission, the output states have energies at the energy eigenvalues. With the double slit experiment, if detectors are placed over the slits, the possible outputs are the detection of passage by the particle through one slit or the other. With removal of the detectors the output particles travel in particular directions with probabilities determined by the interference pattern. The outcome of a measurement will be an eigenvalue if the input is in an output eigenstate. However when the input is stochastic, the output is stochastic and the average result of a large number of measurements is the expectation value. On each occasion, as our examples illustrate, the output particle will be observed as having a particular position or direction trajectory or energy, and the system description changes from a stochastic description to that for a specific actual system. Thus ψ_{P2} changes to a delta function at the observed direction or energy, etc., and ψ_{P2} can be removed from the description.

12.5 The postulates of quantum mechanics

We are now in a position to set quantum mechanics on a formal basis within the new theory. Remember that our task is either to derive the postulates of quantum mechanics in their entirety, or to show to what

The origin of probability in quantum mechanics

extent they can be derived. We are not assuming the postulates in our theory, and neither are we suggesting that they have a significance other than what we derive within our theory. We follow the scheme of postulates set out by Dicke and Wittke (1960). They are listed in Appendix F. The significance of each postulate in the context of the new theory is examined.

Two cases are introduced. Case 1 concerns the translational wave function ψ_T and the well wave function ψ_W. These are wave functions which distribute the particle gravitational mass and charge densities as described in Chapter 4. ψ_T and ψ_W describe actual clouds. Just as the particle is not a point, but a volume filled with charge and gravitational mass, then ψ_T and ψ_W distribute the oscillatory charge and gravitational mass over a larger volume. Associated with these distributions are actual steady state and gravitational mass densities which integrate to the particle charge and gravitational mass respectively. Case 2 concerns ψ_P which introduces probability. The detail is in Appendix F. The conclusions from the Appendix are summarized in Table 12.1.

There are mathematical considerations which are shown in the left hand columns, and there are matters of interpretation shown in the right hand columns. Turning attention to the Case 1 column in the table, it can be seen that Postulates 2, 3, 4, 6 and 7, which concern mathematical aspects, apply. Postulate 5 which concerns probability does not apply to Case 1.

In Chapter 4 we derived the principles of classical mechanics, and we developed quantum mechanics alongside it. Thus quantum mechanics is not more fundamental than classical mechanics. Quantum mechanics can describe situations which classical mechanics cannot, where distribution is dominant, and so we have wave packets in scattering. However the conservation of energy and inertial momentum is over-riding and indeed this is rigorously applied in QFT. So in scattering, definite energy and inertial momentum values are required for the individual input and output states within the stochastic ensembles. The quantum mechanical momentum operators are applied to the ψ_T's and not to ψ_P. So Postulates 6 and 7 apply to ψ_T and also to ψ_W but do not apply to ψ_P. Hence in Table 12.1 we see that Postulates 2, 3, 4 and 5 apply to Case 2. Our theory differs from conventional quantum mechanics in the physical interpretation to be attached to any one postulate.

12.5 The postulates of quantum mechanics

Table 12.1 Summary of the extent to which the Postulates of Quantum Mechanics follow from the new theory

Post-ulate	Mathematics		Interpretation	
	Case 1 ψ_T ψ_W	Case 2 ψ_P	The New Theory	The Postulates
1			ψ_T, ψ_W and ψ_P do not determine all that can be known about the system	The wave function determines all that can be known about the system.
2	Applies	Applies	Treat measurement as a process	
3	Applies	Applies		
4	The momentum eigenfunctions ψ_T form a complete set	The ψ_T form a complete set each labelled by ξ and weighted by $\psi_P(\xi)$		
5			For the set of $\psi_P\psi_T$ or $\psi_P\psi_W$ wave functions $|c_i|^2 = |\psi_{Pi}|^2$ where $|\psi_{Pi}|^2$ is the probability of finding the system in the state i	$|c_i|^2$ is the probability of finding the system in the state i
6	Applies with a caveat – see section F.7			
7	Applies			

The origin of probability in quantum mechanics

Nevertheless there is sufficient agreement shown under the heading 'mathematics' that the mathematics which stems from our theory and from the conventional postulates is identical. Specifically, by inspection, the mathematics in Dicke and Wittke (1960) follows from our new theory.

12.6 Summary

Probability is introduced via the stochastic factor ψ_P converting quantum mechanics into a form of statistical mechanics. ψ_P has been applied to Rutherford scattering, spontaneous emission and the double slit experiment. The introduction of probability allows the establishing of quantum mechanics formally within the new theory. This is achieved by an examination of the postulates of quantum theory and the demonstration that they can be derived from the new theory. There are differences in interpretation between the new theory and conventional quantum mechanics. Nevertheless the mathematics of quantum mechanics follows from the new theory. We have put quantum mechanics on a formal basis within the new theory and this provides the context for the examination of QFT in the next chapter.

Chapter 12 references

Al-Khalili J 2012 *Quantum A guide for the perplexed* (Phoenix)
Baggott J 2011 *The quantum story* (Oxford)
Cox B and Forshaw J 2012 *The quantum universe: everything that can happen does happen* (Penguin Books)
Das A and Ferbel T 1994 *Introduction to nuclear and particle physics* (John Wiley)
Dicke R H and Wittke J P 1960 *Introduction to quantum mechanics* (Addison-Wesley)
Goldstein H 1959 *Classical mechanics* (Addison-Wesley)
Kochen S and Specker E P 1967 J Math Mech **17** 59
Klyachko A A et al arXiv: 0706.0126v4 [quant-ph] 15 Jul 2008
Kumar M 2008 *Quantum* (Icon Books)
Lapkiewicz R et al arXiv: 1106.448 [quant-ph] 22 Jun 2011
Penrose R 2005 *The road to reality* (Vintage)

Chapter 13

Quantum field theory and quantum electrodynamics

13.1 Introduction

Quantum field theory (QFT) is a major extension of quantum mechanics and is principally applied to particle decays and particle interactions. The essentials of each of these processes are then captured in a Feynman diagram. QFT is a general theory and quantum electrodynamics (QED) is the area of its application to electrons and other charged particles which interact electromagnetically. The purpose of this chapter is to demonstrate that quantum field theory and electrodynamics follow from our new theory. We use Weinberg's book 'The quantum theory of fields', Volume 1 (Weinberg 1996) as a definitive account and we delve into appropriate detail below.

13.2 Relativistic quantum mechanics

We have already shown in Chapter 4 how Schrödinger's equation arises from the new theory, and in the previous chapter and Appendix F we show formally how non-relativistic quantum mechanics arises from the new theory. Our first step here is to establish that relativistic quantum mechanics follows from our theory. This is dealt with in Chapter 12 in the Book along the following lines. We show in Chapter 4 that

$$\omega^2 - c^2 k^2 = \omega_0^2$$

Using from Chapter 12 and Appendix F, where $\mathcal{H}$ is the Hamiltonian operator, $\mathcal{P}$ is the inertial momentum operator and ψ_T is a translational wave function,

$$\mathcal{H}\psi_T = -i\hbar \frac{\partial \psi_T}{\partial t}$$

Quantum field theory and quantum electrodynamics

$$\mathcal{P}\psi_T = i\hbar \frac{\partial \psi_T}{\partial z} \mathbf{z}_1$$

where $\mathcal{M}_0$ is the particle inertial rest mass and $\mathbf{z}_1$ is the unit vector along z. With the local z axis slowly changing direction in space and time we obtain

$$\mathcal{H}^2 \psi_T = (\mathcal{P}^2 + \mathcal{M}_0^2)\psi_T$$

(13.1)

This is the Klein – Gordon equation and was used by Dirac to obtain his linearized equation form. These are the steps that show that Dirac relativistic quantum mechanics follows from the new theory.

13.3 Quantum mechanics and the basis for quantum field theory

The previous chapter shows that the postulates of quantum mechanics follow from our new theory, with some caveats with regard to interpretation, and the section above shows how the relativistic Dirac quantum mechanics follows from our theory. We can now highlight the following features of conventional quantum theory which also follow from the new theory.

Matrix representations. Wave functions, including the factor ψ_P, can be expressed as a linear combination of basis functions. The set of coefficients in the linear expansion can be set out in the form of a column matrix and this description is known as a matrix representation of the quantum mechanical state. There are various ways in which matrix representations can be made, and they include the Schrödinger, Heisenberg and interaction representations, and their definitions can be found in quantum mechanics textbooks.

Hilbert space. The concept of a representation leads to the wave function, including ψ_P, to be considered as a state vector in a suitable vector space, called a Hilbert space.

Probability. In the previous chapter the $\psi_P \psi_D$ construction (where ψ_D equals $\psi_T \psi_R$ or ψ_W in line with the wave functions in Chapter 4 of the Book) introduces probability and it is shown that the $|c_i|^2$ provide the transition probabilities between input and output states, see Table 12.1. $\sum_i |c_i|^2$ over all allowed output states is unity

QFT is based on conventional quantum theory and its relativistic Dirac extension. In his Chapter 2 pp 49-50 Weinberg sets out the

13.4 Particle description, symmetries and the standard model

specific features that are required:
(1) Hilbert space and rays in Hilbert space as descriptions of quantum states
(2) Operators are Hermitian
(3) Weinberg equation 2.1.7, page 50, states that the transition probability from ray R to ray R_n is $|(\Psi, \Psi_n)|^2$ where Ψ and Ψ_n are any vectors belonging to R and R_n. Our $|c_i|^2$ result (see Appendix F, Section F6) is consistent with this. A further feature is the extension to scattering with many particle input and output. Weinberg in the introduction to his Chapter 3 sets out the probabilistic nature of these interactions and our introduction of probability into the scattering process in the previous chapter is consistent with his picture.

A further and major feature in QFT is the conservation of energy and inertial momentum. This comes straight from the relativistic form of classical mechanics. In our theory the particle energy and inertial momentum are well defined by ψ_T via the angular frequency ω and wave vector k consistent with special relativity. Energy and inertial momentum conservation, and of course charge conservation, apply.

QFT is developed to determine transition rates and cross sections from the probability $|(\Psi, \Psi_n)|^2$, and to develop the concept of Feynman diagrams, which we will introduce later via examples. Note the conclusion in Chapter 11 that positive gravitational mass anti-particles are the case in our universe. As a consequence Feynman diagrams, which are constructed on the basis of the conservation of the energy and inertial momentum, automatically satisfy gravitational mass and gravitational momentum conservation. We conclude that the features required for developing QFT are in quantum mechanics as derived from our theory.

13.4 Particle description, symmetries and the standard model

The discussion above leads on to the following point regarding the scope of QFT. We are interpreting QFT as a general theory and which includes the treatment of fundamental symmetries, one of which is the effect of the Lorentz transformation. Having established the general theory, it can be applied to cases in which specific particle properties and various symmetries (or conservation rules) are introduced. Symmetries are examined by looking for conservation of a property

Quantum field theory and quantum electrodynamics

having applied an operator.

Three such operators are conversion of a system from particles to anti-particles (C), parity (P) which is a spatial inversion (the definition of parity in the new theory requires careful examination), and time reversal (T) in which t is replaced by $-t$. These have not been examined in the context of the new theory and it is not known what the new theory will predict. The experimental results of the full consequences of these operators is not yet complete. Theoretically with each process the various conservation principles and symmetries are applied so that the QFT In and Out states are descriptions of equivalent systems with regard to the conserved entities and are therefore treated as descriptions of different states of the same system. For example in the next chapter we use the principle of gravitational mass and charge configuration matching so that the three dimensional pattern of the gravitational mass and charge of the output basic particle(s) matches the gravitational mass and charge pattern of the input basic particle(s). Given this match, the computational issue for QFT concerns the extent of matching of the quantum mechanical description of the output state to that of the input state.

Conventionally the standard model is the source of particle properties and symmetries (e.g. baryon conservation, lepton type conservation) whereas the new theory has to provide the particle properties and symmetries. In the new theory the gravitational mass and charge densities are obtained using the convolution of the wave function with the particle description. When we drop the convolution, retaining the wave function only, we arrive at the quantum mechanical description. QFT uses the wave function corresponding to $\psi_P \psi_D$ and the ramifications of the representations discussed above. Thus in using QFT in the context of the new theory, the $\psi_P \psi_D \psi_P^* \psi_D^*$ description replaces $\psi_P \psi_D \psi_P^* \psi_D^* * \rho_B$ and $\psi_P \psi_D \psi_P^* \psi_D^* * m_B$. This is equivalent to replacing our particle models by point particles This means that particle properties are to be separately input to QFT, just as they are conventionally using the standard model. In the new theory it is from the particle descriptions that we obtain the various conservation results and symmetries and these also are to be separately input to QFT. In parallel with making these inputs to QFT, the new theory faces the challenge of either confirming the details of the standard model or providing alternatives for both particle properties and symmetries. This is discussed in Chapters 14 and 15.

13.5 Quantum electrodynamics

The conclusion from the preceding section is that the starting point for Weinberg's exposition follows from our theory. However we need to attend to the points above regarding particle properties and symmetries in order to check that the details of the treatment of quantum field theory in Weinberg Volume 1 follows from our theory. This amounts to checking some of the details of the application of QFT to QED, which we do in the following sections.

13.5 Quantum electrodynamics

Weinberg, Chapters 3 to 7, sets out the general principles of QFT, which follow from our theory, and the principles are applied to electrodynamics in Chapter 8 et seq. We consider three major QED constituents: the virtual photon propagator, vertex loops and vacuum ionization e^+e^- loops.

Weinberg's starting point is to assume gauge invariance (to account for massless particles with spin, page 339) rather than assuming Maxwell's equations, and the analysis leads to the Hamiltonian, equation (8.4.1), on page 350, and thence on to the construction of Feynman diagrams. The Hamiltonian has been derived using the Coulomb gauge which is not Lorentz invariant, and the approach necessitates coping with this problem. An alternative (Weinberg page 345) is to use the BRST formalism discussed on pages 35-36 Weinberg Volume 2 which allows freedom from dependence on Lorentz-non-invariant gauges in obtaining canonical quantisation.

In our theory in Chapter 1 of the Book we have already derived Maxwell's equations. We can account for the origin of the potentials via the omega waveforms, and we have a model for photons and hence a scheme of quantisation. The energy transported by a photon has been obtained for plane polarisation, see Section 3A.11 in the Book. A photon with $\pm\hbar$ angular momentum can be constructed using circular polarisation. Our magnetic vector potential **A** is defined in Chapter 1 of the Book such that

$$\frac{\partial V}{\partial t} + c^2 \operatorname{div} \mathbf{A} = 0$$

In conventional QED parlance this is the Lorentz gauge, and of course it is Lorentz invariant. In Appendix 4C of the Book we present an explanation for the physical origin of a particle's canonical

Quantum field theory and quantum electrodynamics

momentum. This leads to the relativistic form of the classical force involving the magnetic vector potential, used in Section 4.7 of the Book. It is a standard piece of analysis to then obtain the relativistic Lagrangian and the relativistic Hamiltonian in canonical form, for example see McCrea (1954 pp56, 57), and then to add the contribution from the photon. Inspection of the Weinberg expression for the Hamiltonian shows that this follows from our theory and therefore the edifice of Feynman diagrams then follows. An example of a Feynman diagram is shown in Figure 13.1(a) for $e^+e^- \to \mu^+\mu^-$ and the conventional QED treatment of this process follows from the new theory.

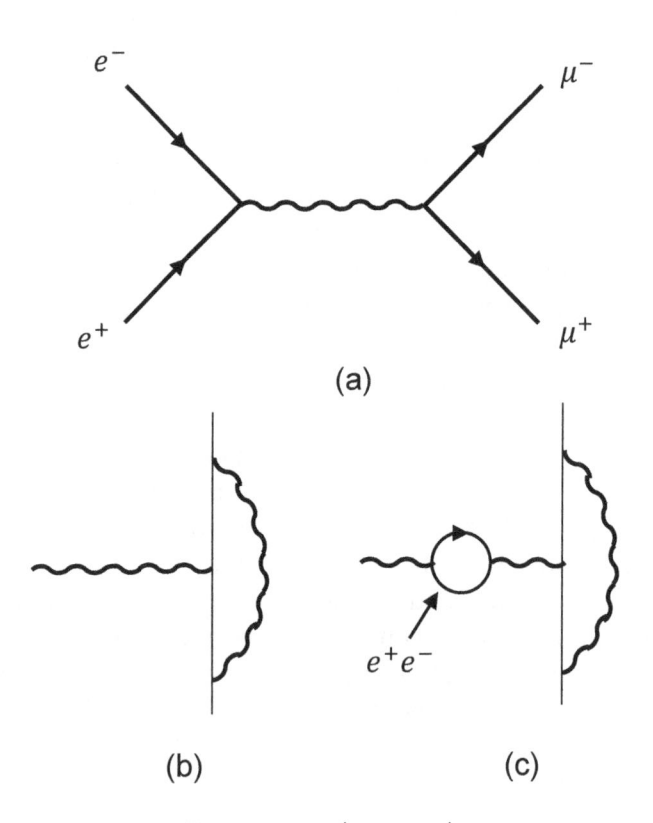

Figure 13.1 Feynman diagrams (a)$e^+e^- \to \mu^+\mu^-$. The wavy line represents a photon (b) single electron with a photon loop around the vertex. The straight line represents the electron (c) Modification of the previous diagram by the introduction of an e^+e^- loop into the external photon line.

The diagram is an outcome of the theoretical treatment. The input

13.6 QED treatment of the Lamb shift

vertex is connected to the output vertex by the virtual photon propagator. The propagator has associated with it a factor from the theory. The physics has already been input to the theory as described above, and so we are content to go along with whatever name is attached to it, in this case the 'virtual photon propagator'.

Figure 13.1(b) is the diagram for an electron moving in static external magnetic and electric fields and again this follows from our theory. The photon loop connects the incoming and outgoing electron and modifies the current at the vertex. In Chapter 8, Section 8.8, we show that the magnetic moment of the electron is $\mu_0 e\hbar/2\mathcal{M}_{0e}$ and this is consistent with the requirement from Dirac quantum mechanics. Conventional QED assumes this value. The Feynman diagram in Figure 13.1(b) is used as the starting point for the calculation of QED corrections to the electron magnetic moment, or more strictly, the gyromagnetic ratio. Comprehensive calculations lead to corrections of a few parts in 10^{10} or better and these remarkable predictions are confirmed by equally remarkable precision measurements, for example see Baggott (2011) page 191 for details.

13.6 QED treatment of the Lamb shift

In this section we pick out some features from the QED treatment of the Lamb shift. The Lamb shift concerns the observation of a separation of two energy levels in the hydrogen atom giving rise to an absorption at around 1 GHz. The QED analysis is able to obtain agreement with experiment within about 10 kHz, Weinberg p594. In the next section we examine the consequences of using our model of the electron in this context. The conventional QED model of the electron consists of a bare negative charge surrounded by positrons. This is based on the following. An e^+e^- pair is formed from ionization of the vacuum. The bare negative charge of the electron attracts the positrons and repels their electron companions, so that the bare charge is partially shielded, resulting in the electron having an effective charge of $-e$. In the vicinity of the proton the positrons are repelled changing the effective charge to the bare charge $-e_B$, i.e. as the electron moves, the nucleus penetrates the outer positrons. The bare charge has a charge magnitude greater than e and this is consistent with the outcome of the QED analysis when the bare charge plus positrons is polarised in the vicinity of the proton. So the electron becomes a

Quantum field theory and quantum electrodynamics

many particle system and the electron cloud has a radius a given by Weinberg (page 493),

$$a^2 = -\frac{FSC}{\pi}\left(\frac{\hbar^2}{c^2\mathcal{M}_{0e}^2}\right)\left[\ln\left(\frac{\mu^2}{\mathcal{M}_{0e}^2}\right) + \frac{2}{5} + \frac{3}{4}\right]$$

where the equation has been recast in SI units, FSC is the fine structure constant, $\mathcal{M}_{0e}$ is the electron inertial rest mass and μ is an infrared cutoff with the value 16.64 Ry $\equiv$ 226 eV with 1Ry = 13.6057 eV (Weinberg page 593), leading to a radius of around 70 fm. This is applicable to the electron in the vicinity of a nucleus and therefore applies in the case of the treatment of the Lamb shift.

13.7 The new theory's electron model in the treatment of the Lamb shift

The new theory brings an electron model which arises outside of and is more fundamental than QFT and QED. Chapter 7 in the Book concluded that the electron has a ring structure shown in Figures 7.2 and 8.8, and reproduced in Figure 7.1 in this book. We take at face value the physical properties required by QED for the electron in the proximity of a nucleus. In considering the origin of the Lamb shift we have to reconcile our electron model with e^+e^- pair creation and the size of the electron cloud.

The main objection to electron-positron pairs popping in and out of existence as a consequence of the uncertainty principle is that gravitational mass and energy are not conserved. Our theory certainly allows for e^+e^- vacuum polarisation in that there can be creation of e^+e^- pairs given sufficient energy and gravitational mass. The key to this process is to allow the production of fractional charge e^+e^- pairs. There is then no difficulty with the supply and release of gravitational mass and energy. These positrons and electrons are not in free space interacting with universal photons and so the condition derived in Chapter 3 that the charge magnitude is e does not apply. We can now introduce the e^+e^- loop as shown in Figure 13.1(c).

However this leaves the issue that the electron ring size (385 fm radius) is large compared to the 70 fm radius sphere predicted by conventional QED. So allow the electron to reduce its size to the radius required in the vicinity of the nucleus. In so doing the electron reduces its electrical potential energy while keeping the inertial and gravitational rest masses constant, but subject to the minor variations for the whole electron system required by QED. Similarly we expect that in the vicinity of the nucleus the e^+e^- pairs can be formed with smaller ring sizes. So with an e^- ring becoming about 70 fm radius near to the nucleus and in combination with e^+e^- pairs, a cloud around 70 fm radius can be generated.

We therefore take it that our particle behaves exactly as in conventional QED. The electron is decorated with e^+e^- pairs, and e^+e^- pairs generated by vacuum polarisation are shown as loops in propagators in Feynman diagrams. There is charge screening and then exposure of the bare charge in the vicinity of the nucleus. So we conclude that our electron model can be consistent with the QED electron model and the QED treatment of the Lamb shift.

13.8 Conclusions

The starting point for Weinberg's exposition follows from our theory and the development of the fundamentals of QFT then follow. Following the discussion above, QED follows from the new theory. We conclude that the whole treatment of quantum field theory (with the exception of the treatment of the symmetry transformations CPT), together with quantum electrodynamics, in Weinberg Volume 1 follows from our theory.

13.9 Summary

It has been demonstrated that the fundamentals of quantum field theory (QFT) follow from the new theory, which importantly allows the use of Feynman diagrams. In so doing the new theory's construction of a convolution of a wave function and a particle function is approximated by a wave function only. This means that particle properties and the resulting symmetries are to be input to QFT separately. Quantum electrodynamics (QED) is the application of QFT to the behaviour of charged particles in electromagnetic fields. The conventional QED results follow from the new theory and these include the QED predictions regarding the corrections to the electron magnetic moment and the value of the Lamb shift.

Quantum field theory and quantum electrodynamics

Chapter 13 references

Baggott J 2011 *The quantum story* Oxford University Press
McCrea W H 1954 *Relativity physics* Methuen
Weinberg S 1996 *The quantum theory of fields* Volume 1 and Volume 2, Cambridge University Press

Chapter 14
Composite particles

14.1 Introduction

In Chapter 2 we introduced what we call the single omega solution for fundamental particles. In this chapter we extend the stable of particles with the introduction of the composite particle. It is composed of independent Dirac particles, and we apply the concept to account for the excited states of the proton and the structures of kaons and pions. A composite particle can provide either a stable particle or a precursor structure enroute to a stable particle. A tool in the analysis of composite particles is the x – energy diagram and this is where we start.

14.2 The x – energy diagram and the nucleon single omega solutions

In Chapter 2 we have found solutions for the particle internal oscillatory fields. They are classified according to the value of the spin parameter s. $s = \frac{1}{2}$ solutions are used for the proton and neutron in Chapter 6 and for the electron and muon in Chapter 7. $s = 0$ is used for the kaon and pion in Chapter 6. In all these cases the angular frequency is the same for all parts of the particle. We refer to these particles as single omega solution particles. These are distinct from composite particles, introduced below, where different parts of the particle have different frequencies.

A key tool in what follows is the x – energy diagram. In this section we demonstrate its use with a single omega solution for the proton. The diagram in Figure 14.1 has the particle energy (proportional to the angular frequency ω) along the horizontal axis and the x value of either cylindrical (radius r_c) or spherical (radius r) shells along the vertical axis.

Composite particles

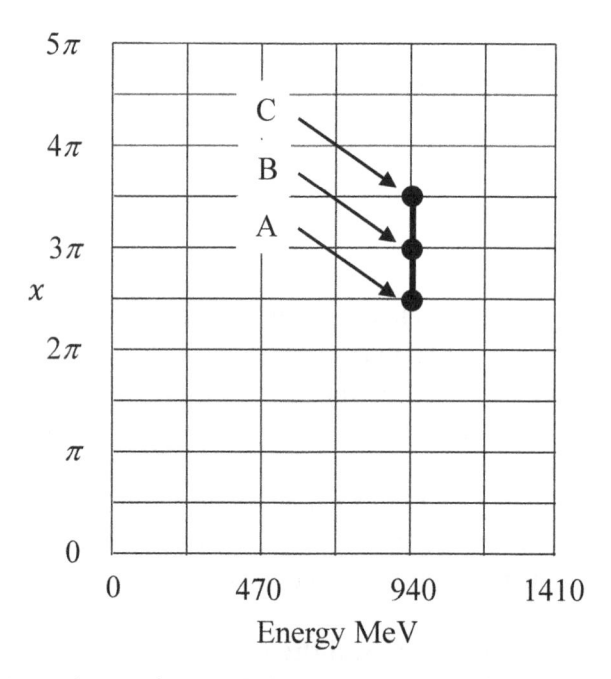

Figure 14.1 The x values at A (7.85), B (9.42) and C (11.00) using Set 3 from Chapter 9 for the proton shell centres for the single omega solution

Thus x equals $r_c\omega/c$ or $r\omega/c$. The proton energy is 938.3 MeV and is approximated here by 940 MeV for the purposes of illustration and consistency with later diagrams. The centre radii of the cylindrical shells are at integer multiples of $\pi/2$, see Chapter 6, Section 6.5. The proton shell centre radii, x_{0i}, are at 7.85, 9.42 and 11.00 for Set 3 (see Chapter 9, Section 9.2) indicated by A, B and C on the diagram. A similar diagram applies for the neutron single omega solution.

14.3 The x – energy diagram and independent Dirac particles

In this section we examine how a single omega solution particle, given sufficient energy, can convert to a composite particle in which each shell acts as an independent entity. On Figure 14.2 we have indicated, with a square, a shell constituent of a cylindrical single omega solution particle with energy E. Along the straight line joining the centre of the square to the origin, the centre radius $r_c = xc/\omega$ is constant where $\hbar\omega = E$.

14.3 The x – energy diagram and independent Dirac particles

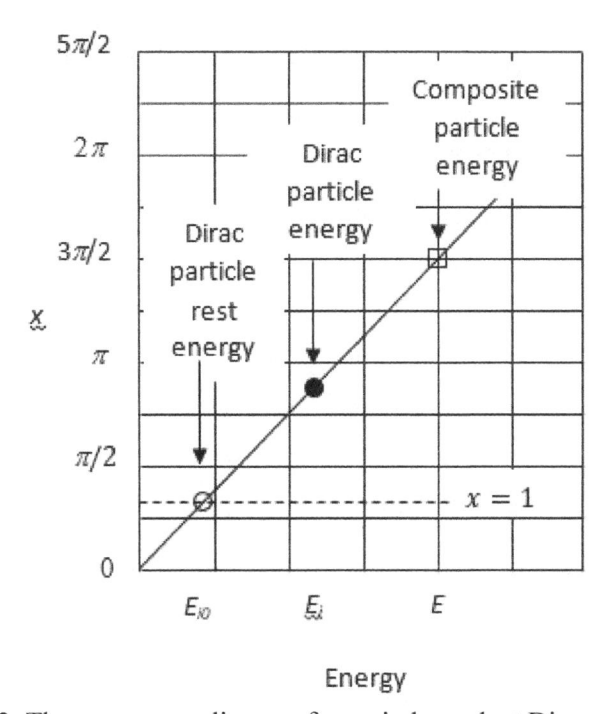

Figure 14.2 The x – energy diagram for an independent Dirac particle showing the particle rest energy (open circle), the particle energy (filled circle), and the x value indicated by a square at the composite particle energy

When $x = 1$ the energy is $E_{i0} = \hbar\omega_{i0}$ and the radius can be expressed as $r_{ci} = c/\omega_{i0}$ where the subscript i denotes the ith Dirac particle. The magnetic moment of a particle confined to a circle with radius r_{ci} is

$$\mu_0 i_i \pi r_{ci}^2 = \frac{\mu_0 q \hbar}{2\mathcal{M}_{i0}}$$

where i_i is the current, $\mathcal{M}_{i0}c^2 = E_{i0}$ and q is the charge which is not necessarily the electronic charge. This is the Dirac magnetic moment for a particle with spin ½ provided that $\mathcal{M}_{i0}$ is the Dirac particle rest mass.

So let's start with the Dirac particle corresponding to a single omega solution at the rest energy E_{i0}. Confine the particle to a delta function in z. This ensures it has a cylindrical type solution since single omega solutions with $\theta = \pi/2$ in equations (6.2) and (6.3) in the Book are functions of r_c, multiplied by a spin function. The lower

Composite particles

edge of the shell at E_{i0} cannot be below $x = \pi/4$ because this is the point at which $x' = x - \pi/4$ becomes zero, where x', the parameter of the associated spherical shell (see Section 6.4), becomes zero. Let's increase the x extent of the particle to the maximum $2\Delta x = 2(1 - \pi/4) = 0.43$ which takes it down to the edge of the shell and up to an x value which keeps the magnetic moment at the Dirac value, and so the particle is described by an annulus shown in Figure 14.3(a).

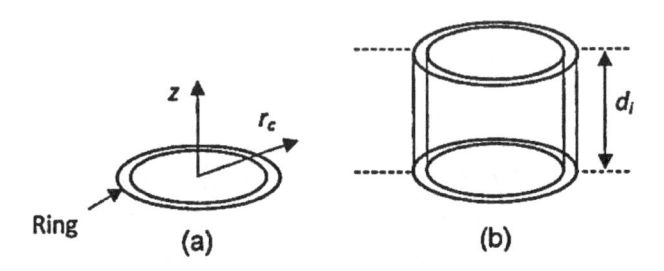

Figure 14.3 (a) Basic ring structure for a Dirac particle (b) the distributed ring filling a cylindrical shell for a Dirac particle in motion

At rest the oscillatory solution has an angular frequency corresponding to that of the rest energy and, together with the steady state solution, these constitute a single omega solution within one shell. It has a gravitational mass corresponding to the rest energy and it can have any charge magnitude, i.e. it is not constrained to $\pm e$ because it is not in free space.

Let's give energy to the particle by motion along z with a wave function ψ_T, resulting in the volume shown on Figure 14.3(b) and indicated on Figure 14.2 by the filled circle at energy E_i. Since a fractional e charged particle cannot have an independent existence, the motion will be described by a standing wave function within the composite particle. The wave function will be sinusoidal, and so we arrive at the situation of Figure 14.4 where $\psi_T = \cos k_i z$. Inspection of $\psi_T \psi_T^*$ shows that most of the Dirac particle is confined to the length d_i where $k_i = \pi/2d_i$. So the Dirac ring particle is distributed and becomes a cylindrical particle. It approximates to a shell filled to a length d_i along z. Appendix D, Section D.2 discusses the various issues to do with cylindrical and partial sphere approximations.

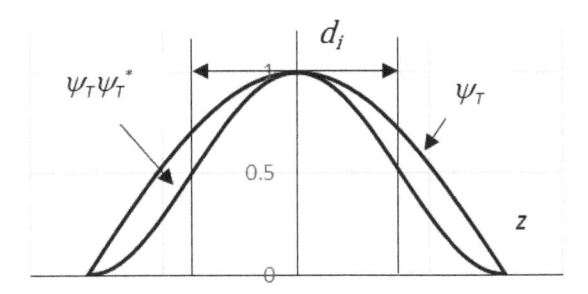

Figure 14.4 The wave function for the Dirac particle

We can construct a composite particle from a number of Dirac particles occupying, or partially occupying, shells at the composite particle energy E. By restricting the x extent of each of the Dirac particles to a maximum of one shell thickness $2\Delta x = \pi/2$ at the composite particle energy, this allows adjacent Dirac particles with differing x_0 values to slide past each other without overlap. Referring to Figure 14.2, when the thickness of the shell at $3\pi/2$ is filled to $\pi/2$, this corresponds to an x thickness of 0.33 at $x = 1$ and is less than the maximum thickness of 0.43 at $x = 1$, and shells with greater x_0 values will have even smaller x thicknesses The Dirac particle's steady state charge and gravitational density distributions correspond to those of a single omega solution shell at its rest mass. In giving motion to the Dirac particle the charge remains the same and equals that of the corresponding single omega solution shell at the composite particle's energy. We show in Appendix G, Section G2, that composite particles are in the sub-set and that a composite particle obeys the sub-set equations (D.8) to (D.11). From Section 13.2, equation (13.1), the Dirac particle energies are given by,

$$\mathcal{M}_i c^2 = E_i = (c^2 \hbar^2 k_i^2 + \mathcal{M}_{i0}^2 c^4)^{1/2}$$

(14.1)

We apply the principles of this section to an analysis of an excited state of the proton in the next section.

Composite particles

14.4 Composite particles and an excited state of the proton

We can now apply the principles above to an excited state of the proton. The details are in Appendix G, Section G.3. The approach here differs from that of Chapter 12 of the Book in three ways (a) the use of rings which are delta functions along z (b) the treatment is applied with individual extents for each shell along z and (c) tails of the wave functions extend beyond the original confines, and as a result we need to rework the analysis which leads to Table 12.1 in the Book. Chapter 12 of the Book introduces the idea that when a proton or neutron becomes excited it becomes a composite particle composed of independent Dirac particles. The associated solution is the original single omega solution and so ρ_{AM}, m_{AM}, charges, total gravitational mass, inertial mass and volume remain the same.

We identify the features of the proton composite particle using the x – energy diagram, Figures 14.5 (a) and (b).

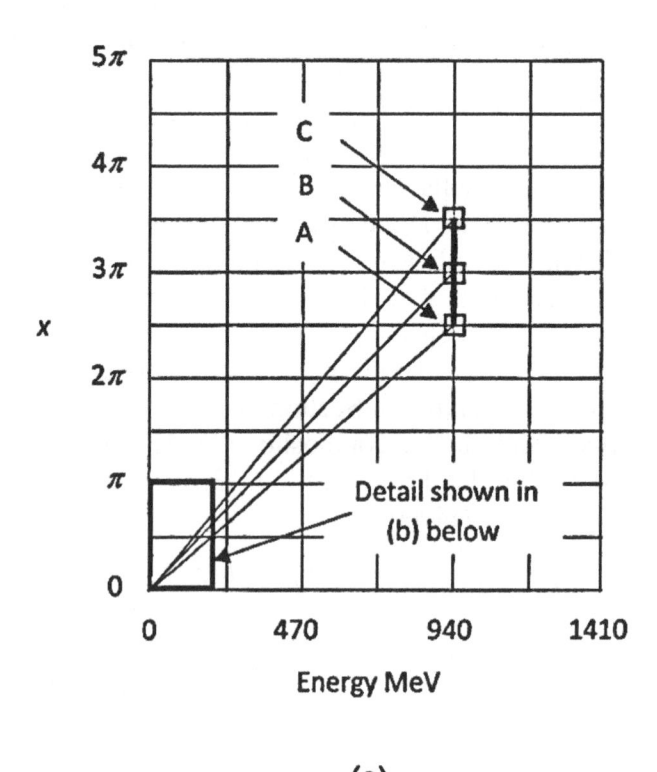

(a)

14.4 Composite particles and an excited state of the proton

(b)

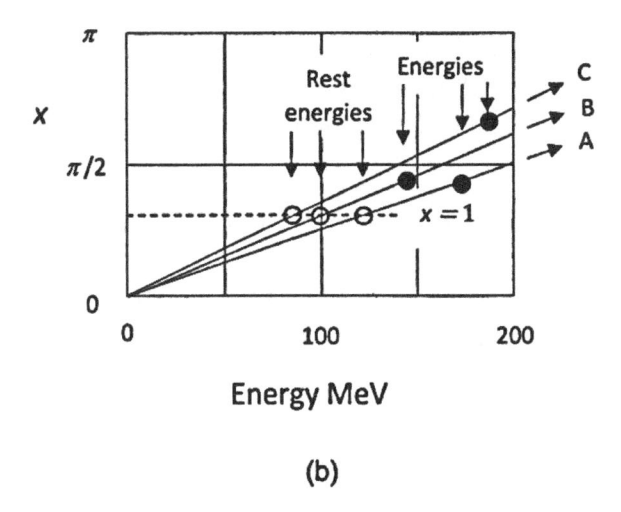

Figure 14.5 The proposed structure for an excited state of the proton as a composite particle composed of three Dirac particles (a) the Dirac particles occupying the shells of the original single omega solution particle (b) the energies and rest energies of the Dirac particles

The rest energies are $938/x_{0i}$ MeV. Instead of a common length we use the d_i for each shell as explained in Section 14.3. Since d_i are known for each shell in the proton and neutron, we can therefore calculate the Dirac particle energies and their sum using (14.1). These are shown in Table 14.1.

Table 14.1 The various parameters for the Dirac particles arising from the bound shells of the proton. Previously Table 12.1 in the Book uses a common extent parameter L. Here the extent of the wave function along z is related to the shell extents d_i along z obtained from Figure 9.4

Shell	x_{0i}	d_i fm	k_i fm^{-1}	$M_{i0}c^2$ MeV	$M_i c^2$ MeV	M_i/M_0
1	7.85	2.50	0.63	120	173	0.18
2	9.42	3.00	0.52	100	144	0.15
3	11.00	1.86	0.84	85	187	0.20
					$\sum_i M_i/M_0$	0.54

Composite particles

The sum as a fraction of the rest energy for the proton is near to 0.5 as previously in Table 12.1 in the Book. We are left with energy $\hbar\omega/2$ not accounted for by the Dirac particles. We account for it with a neutral parton so that the proton energy is given by

$$\mathcal{M}_0 c^2 = \sum_i \hbar\omega_i + \frac{1}{2}\hbar\omega_0 \equiv \hbar\omega_0$$

The neutral parton energy is $\hbar\omega_0/2$. The spin is zero, because there is no polarisation as in a photon. There is no electric charge. A possible description for it is that of an omega waveform in the lowest quantum state, accompanied by gravitational mass of $A\omega/2$. This would mean that the Dirac particles have to account for around $A\omega/2$ gravitational mass, whereas they account for all of the electric charge. The rest energies of the Dirac particles are shown on Figure 14.5 (a). The Dirac particles have their individual energies, also shown on Figure 14.5 (a), and therefore their individual angular frequencies, in distinction from the common angular frequency of the original single omega solution for the proton shown in Figure 14.1. The charge fractions of the Dirac particles are the charge fractions shown on Figure 9.4, i.e. 0.68 at 7.85, 0.68 at 9.42 and -0.35 at 11.00.

We expect that there should be higher energy states described by wave functions containing more zero crossings both for the proton and for the neutron. These excited states are not investigated further. So there are no predictions yet which can be compared with the experimental data. In fact it is a convenient assumption that the energy of the lowest excited state is that of the original particle at around 940 MeV, whereas in a rigorous treatment we would predict the precise energy of this state. However in one key respect the model of protons and neutrons composed of the Dirac particles together with a neutral component is in agreement with the major conclusions from electron-nucleon scattering experiments and allows us to identify the neutral component as the neutral parton. An account of these experiments can be found in text books e.g. Cottingham and Greenwood (2007).

14.5 Kaons

In Chapter 6 we propose that the kaon is a single omega solution particle with spin of zero and is composed of spherical shells. In the

14.5 Kaons

Book in Chapter 6 and followed up in Chapter 8 in the Book we predict a particle rest energy of 330 MeV for both the charged and neutral forms. This energy is much less than the observed values of 494 MeV for the charged kaons and 498 MeV for the neutral kaons (Particle Data Group 2014). In Chapter 12 of the Book we predict a revised value of 527 MeV for a cylindrical form with two shells and it is proposed that it converts to the spherical single omega solution form with the same energy and that these forms have different lifetimes.

Here we develop these ideas using the concept of the composite particle developed above. We deal with charged kaons first. It is proposed that potentially the charged kaons have two forms, the cylindrical and the spherical, along the lines of Chapter 12 in the Book. The cylindrical forms are composed of independent thin shelled Dirac particles which have x values at the kaon energy of 3.66 and 2.09 (these will be explained shortly) and that the cylindrical form is the precursor of the spherical. Thus, we are considering a situation in which the only way to make the single omega solution form of the particle, i.e. the spherical form, is by making the cylindrical form first. Hence the cylindrical form consists of two Dirac particles as shown in Figure 14.6 and we discuss this in detail below.

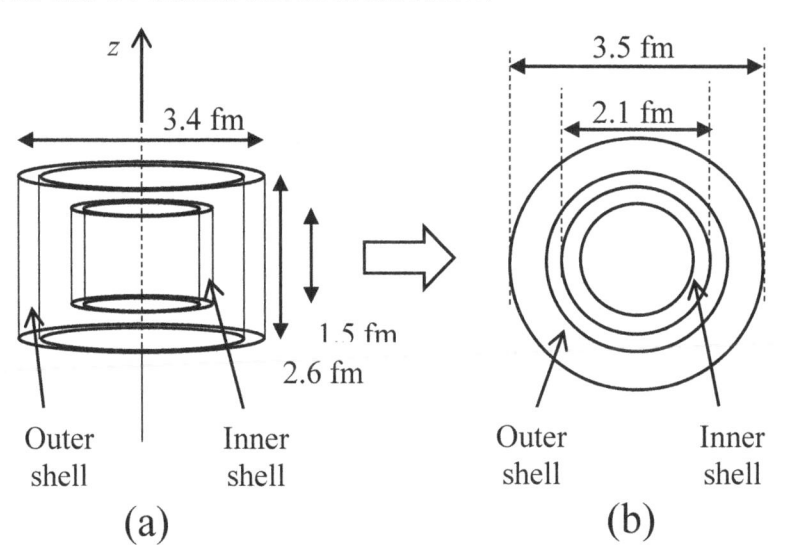

Figure 14.6 Forms of the charged kaons (a) the cylindrical form with two shells (b) the spherical form with two shells. The cylindrical form is the precursor of the spherical form

Composite particles

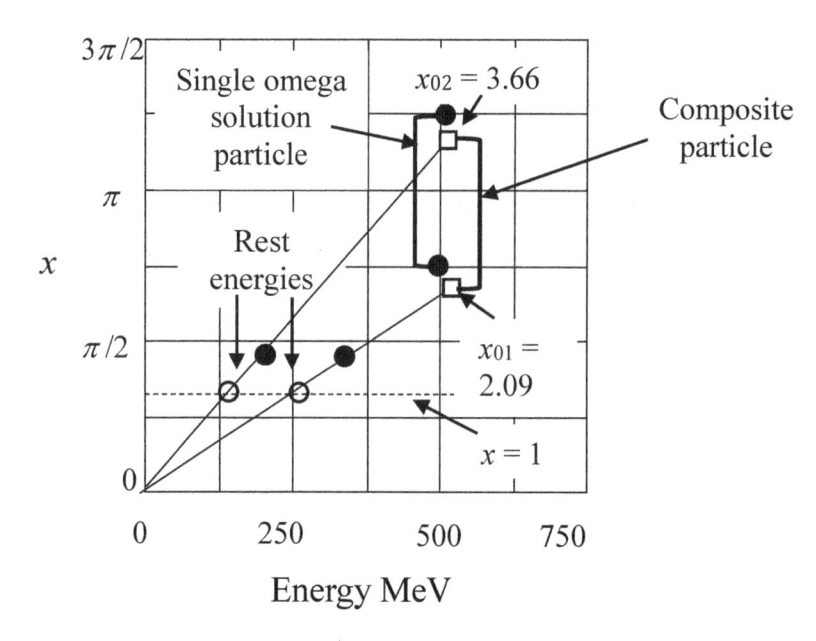

Figure 14.7 The charged kaons, K$^{\pm}$. The lower box is centred at x_{01} and the upper is centred at x_{02} signifying occupation of shells in the cylindrical composite particle at energy 507 MeV. The lines to the origin represent the independent Dirac particle constituents of the composite particle with energies at 323 MeV and 184 MeV, shown by the filled circles, and rest energies indicated by the open circles. The composite particle converts to the spherical single omega solution particle, also at 507 MeV

So the cylindrical form is formed and converts into the spherical form. The former determines the spherical particle rest energy. We therefore begin with the cylindrical form. The composite particle analysis is in Appendix G, Section G.4, and it uses two principles. The first is that the energy of the particle is the sum of the energies of the Dirac particles. The second principle is that, because the particle is in the sub-set, the volume is proportional to the square root of the particle energy. Since we can determine the geometrical configuration we can write down an expression for the volume. The results of the analysis are as follows. The outer shell is centred on $x_{02} = 3.66$ with a filled shell, and the inner shell is centred on $x_{01} = 2.09$ and is partially filled with $2\Delta x = 0.9$, see Figure 14.6. The charge fractions are 0.64 in the outer shell and 0.36 in the inner shell. The Dirac particle energies, 323 MeV in the inner shell and 184 MeV in the outer shell, see Figure 14.7,

14.5 Kaons

sum to the kaon rest energy of 507 MeV which is not far from the measured value of 494 MeV.

With the cylindrical structure a major drawback for stability is, as with the tapered proton model (see Chapter 9, Section 9.1), that it assumes that there is no electric field leakage through the upper and lower surfaces and this clearly is not the case. Figure 14.6 illustrates the conversion from the cylindrical form to the spherical single omega solution. It is concluded below that the spherical form is the predominant form.

We now turn our attention to the neutral kaons. Experimentally it is observed that there are two forms, one with the short lifetime of around 8.95×10^{-11} s and one with the long lifetime of 5.1×10^{-8} s (Particle Data Group 2014). The predominant decay mode of the short lifetime form results in two pions, whereas the predominant decay mode of the long lifetime neutral kaon results in three pions. There is extensive investigation of the neutral kaons in the literature and accounts are given in text books over the years based on the superposition of kaon neutral states (see for example Halzen and Martin (1984 p289), Krane (1988 p692) and Das and Ferbel (1994 p244)).

Again the analysis is in Appendix G. The results are as follows. The cylindrical shells are centred on x values which correspond to those of the spherical particle and therefore ease the transition between the two forms. The outer shell is centred on $x_{02} = 5\pi/4 = 3.9$ and the inner shell is centred on $x_{01} = 3\pi/4 = 2.36$ and both shells are filled, see Figure 14.8. The charge magnitude fractions are equal to 0.53 and the shell charges are of opposite sign. The inner shell energy is 321 MeV and the outer shell energy is 192 MeV and they sum to 513 MeV compared to the measured value of 498 MeV. The conversion from the cylindrical form to the spherical form is illustrated in Figure 14.9.

Composite particles

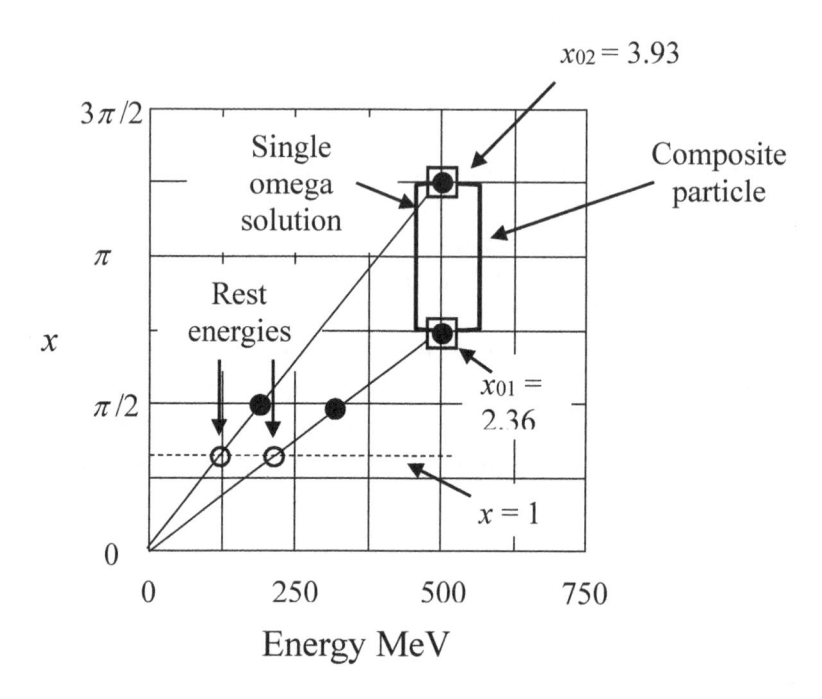

Figure 14.8 The neutral kaon. The lower box is centred at x_{01} and the upper is centred at x_{02} signifying occupation of shells in the cylindrical composite particle at energy 513 MeV. The lines to the origin represent the independent Dirac particle constituents of the composite particle with energies at 321 MeV and 192 MeV, shown by the filled circles, and rest energies indicated by the open circles. The composite particle converts to the spherical single omega solution particle, also at 513 MeV

14.5 Kaons

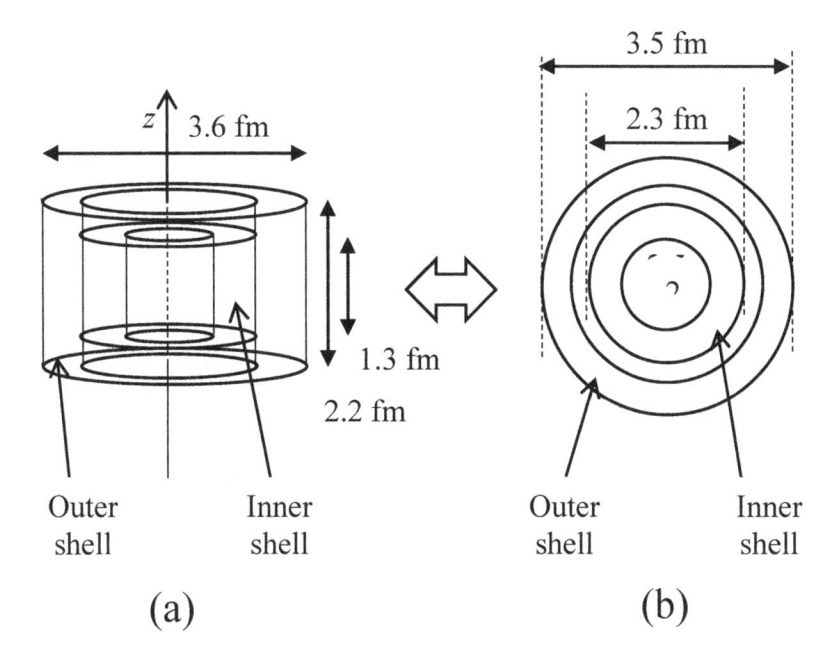

Figure 14.9 Forms of the neutral kaon (a) the cylindrical form with two shells (b) the spherical form with two shells. Both forms are stable and the cylindrical form is the precursor of the spherical form

It is reasonable to suppose that the more likely decay mode is the one in which the initial and final states match each other in their gravitational mass distributions compared to decay modes with a greater disparity in gravitational mass distributions. A similar result is expected for charge distributions. This is an example of gravitational mass and charge configuration matching. It is easy to see that this can be the case with charged particles but it is not so clear with neutral particles with plus and minus charges. Nevertheless we expect that a predominant decay mode of the cylindrical neutral kaon results in two pions as shown in figure 14.10, and for the spherical kaon to decay into three pions, illustrated in Figure 14.11.

Composite particles

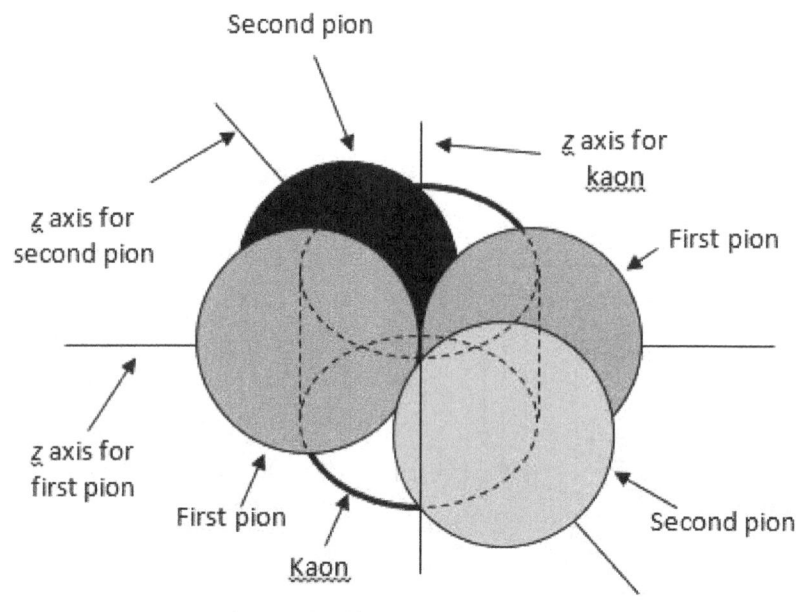

Figure 14.10 Comparison of cylindrical neutral kaon and two neutral pions

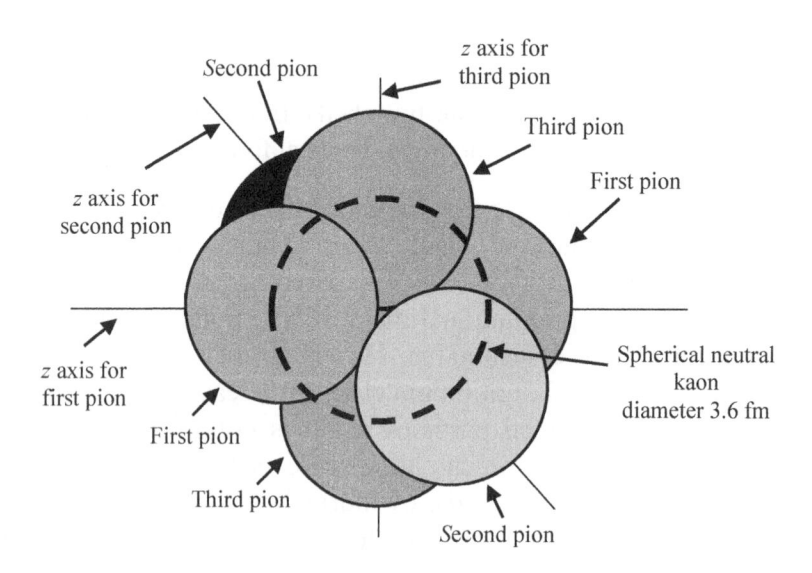

Figure 14.11 Comparison of neutral kaon spherical form with three neutral pions

14.6 Pions

The experimental results allow us to identify the spherical form as the long lifetime kaon and the cylindrical form as the short lifetime kaon.

We return to charged kaons. They are long lived with a lifetime of around 1.238×10^{-8} s (Particle Data Group 2014) but with many decay modes. Krane (1988 p688) lists seven decay modes with 63% into muons, 21% into two pions and 8% into three pions. The Particle Review lists more. We suggest that charged kaons can decay by converting to the cylindrical then via the two pion path, but decay into three pions directly from the spherical form. This is all speculation and requires much more investigation.

14.6 Pions

In Chapter 6 we introduce the single omega solution two sphere model for the charged and neutral pion with a spin of zero, and we calculate the energy in Chapter 8 as 132 MeV with a sphere radius of 1.5 fm. The measured energies are 139.6 MeV for the charged pions and 135.0 MeV for the neutral pion. The section above shows how pions can form from the higher energy kaons, and the pions are in the form of single omega solutions. This prompts the question: can pions be formed also from a composite precursor made of cylindrical independent Dirac particles? We appeal to the gravitational mass and charge configuration matching principle and a possibility is shown in Figure 14.12. The cylindrical shells correspond to x_{0i} at 940 MeV of 6.28 and 7.85 with radii of 1.32 and 1.65 fm giving a mean radius of 1.48 fm. The energy is much greater than the predicted pion energy of 132 MeV. To match this energy would require very much larger radius shells and gravitational mass and charge configuration matching would be destroyed. Note that we have not distinguished which pion we are talking about, i.e. whether charged or neutral. All we can say at this stage is that the charge magnitude fractions for the charged pion have $c_1 + c_2 = 1$ and for the neutral pion $c_1 = c_2$. To develop the analysis further takes us into the topic of the next chapter – the introduction of quarks.

Composite particles

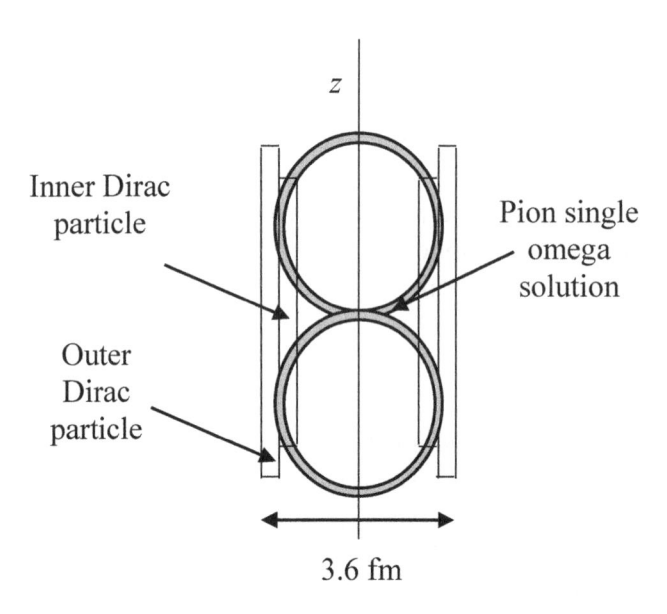

Figure 14.12 The pion single omega solution structure and the cylindrical precursor Dirac particles. The inner particle corresponds to the $x = 6.28$ and the outer particle to the $x = 7.85$ nucleon shells at around 940 MeV

14.7 Summary

Using the tool of the x – energy diagram we have examined the role of independent Dirac particles within particles. This leads to the composite particle structure. The shells of the single omega solution convert to independent Dirac particles in energetic excited states of the proton, and therefore neutrons, along with the neutral parton. Kaons have both spherical single omega solution structures and cylindrical composite structures. The inertial masses of the charged and neutral kaons are predicted, in approximate agreement with their experimental values. An account is given of the structures of the short and long lived versions of the neutral kaon. A composite precursor structure for the pion is proposed.

Chapter 14 references

Chapter 14 references

Cottingham W N and Greenwood D A 2007 *An introduction to the standard model* Second Edition Cambridge

Das A and Ferbel F 1994 *Introduction to nuclear and particle physics* Wiley

Halzen F and Martin A D 1984 *Quarks and Leptons* Wiley

Krane S K 1988 *Introductory Nuclear Physics* Wiley

Particle Data Group, *Review of particle physics*, Chinese Physics C Vol 38, No. 9 (2014) 090001

Chapter 15

Quarks and the structure of hadrons

15.1 Introduction

The number of observed hadrons, whether mesons or baryons, is extensive, see Particle Data Group (2014). The standard model has had considerable success in assigning content to each hadron based on quarks. In this chapter we will show that the quark scheme of the standard model follows from the new theory. This is a bold statement: it will be seen that some of the major principles of the standard model with regard to hadronic physics follow from the new theory, but that considerable divergences remain. Together with probability within quantum mechanics (examined in Chapter 12) and quantum field theory (examined in Chapter 13) this then leads to the treatment of quarks and hadrons within the standard model, and in particular allows us to use the results of the standard model analysis of e^+e^- scattering, the production of muon and quark pairs and the variation with energy of the scattering cross sections. This is dealt with in Sections 15.4 and 15.10. However the development of the new theory does not lead to quantum chromodynamics – instead the overlap potential process introduced in Chapter 10 is a feature of the new theory.

The purpose of this chapter is to set out the new theory's account as to the quark content of particles and the conservation of quarks in particle interactions and decays. In so doing it reproduces much of the standard model's account of quarks. We will not review the standard model here – the reader is referred to text books on the subject. However we summarize at the end of the chapter the similarities and the differences between the new theory's account of quarks and that of the standard model.

In Chapters 9 to 12 of the Book the proposal is developed within the new theory that protons and neutrons are composed of quarks and that these correspond to the up and down quarks of the standard model. In Chapter 12 it is further proposed that one form of the neutral kaon is

15.2 Quarks and the Dirac particle content of nucleons, kaons and pions

composed of two quarks one of which is the strange quark of the standard model. These claims are now developed in detail, and the new theory also leads to the existence of the charm and bottom quarks. A general account of the quark content of hadrons is developed using the concept of the x – energy diagram which was introduced in the previous chapter and which displays the content of particles at particular energies and which can account for the rest mass of quarks.

15.2 Quarks and the Dirac particle content of nucleons, kaons and pions

In the previous chapter we showed how the excited states of the proton and neutron, and the various states of pions and kaons, can be described by structures containing independent Dirac particles. When energy is input to the proton or neutron the shells behave as independent Dirac particles each with its own angular frequency, and each can be distributed by a wave function along the z axis. Our first step here is to review the Dirac particle content of the excited states of the proton and neutron and of the cylindrical forms of pions and kaons. We summarise in Table 15.1 the details of the shell structures of the particles from the previous chapter. The anti-particles have the same dimensions but with the electric charge signs reversed. The x_0 values refer to around 940 MeV except where indicated.

We identify the independent Dirac particles as quarks. Formally we define a quark as follows. It is a Dirac particle of spin ½ and fractional e charge with a particular ring radius r_{c0} (and therefore rest inertial mass) and ring thickness Δr_c so that $2\Delta x = \Delta r_c \omega/c$. Because of the fractional charge it cannot be accelerated by an existing photon and so cannot exist outside a composite particle and is therefore confined to movement within that particle. We interpret Table 15.1 as listing the quark constituents of the various particles.

Quarks and the structure of hadrons

Table 15.1 Summary of the structural parameters and internal charges of composite particles

Particle	Composite form	Shell x_0 at 940 MeV		Shell Thickness $2\Delta x$ at 940 MeV		Charge Units of e
Proton	Excited state	$5\pi/2$ 3π $7\pi/2$	7.85 9.42 11.00	1.57 1.57 1.57		0.68 0.68 -0.35
Neutron	Excited state	$5\pi/2$ 3π $7\pi/2$	7.85 9.42 11.00	1.57 1.57 1.57		-0.32 0.63 -0.32
Pion π^+	Precursor	2π $5\pi/2$	6.28 7.85	1.57 1.57		$c_1 + c_2$ $= 1$
Pion π^0	Precursor	2π $5\pi/2$	6.28 7.85	1.57 1.57		$c_1 = c_2$
Kaon K^+	Precursor	3.87 6.79	(2.09)(1) (3.66)(1)	1.67 2.91	(0.9)(1) (1.57)(1)	0.36 0.64
Kaon K^0	Short and long lifetime forms	4.32 7.20	(2.36)(2) (3.93)(2)	2.88 2.88	(1.57)(2) (1.57)(2)	0.53 0.53

(1) At 507 MeV
(2) At 513 MeV

We now consider a second set of quarks, those around 940 MeV i.e. just above the proton and neutron inertial rest mass energies. These are the quarks indicated on the x-energy diagram Figure 15.1 which have ring thicknesses which fill shells at 940 MeV ($2\Delta x = 1.57$) and shown in Table 15.2.

15.2 Quarks and the Dirac particle content of nucleons, kaons and pions

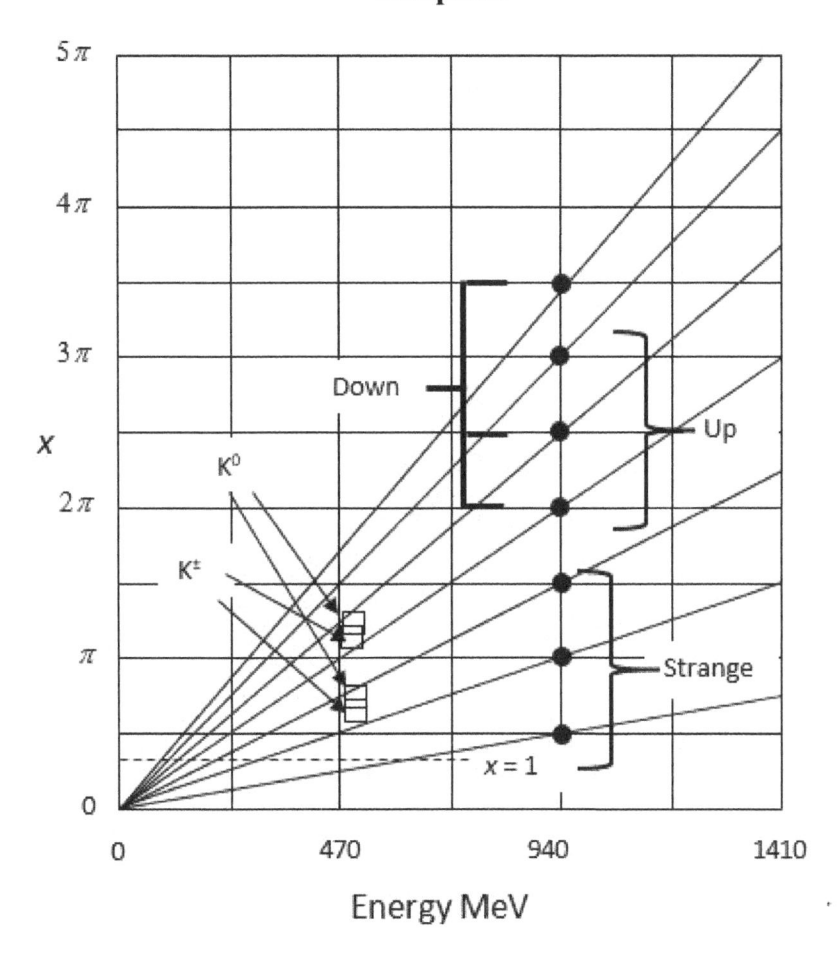

Figure 15.1 The two larger diameter down quarks and the two larger diameter up quarks are those in protons and neutrons. The middle diameter up and down quarks correspond to part of the K^0 content. The smallest diameter up quark corresponds to the part of the $K^\pm$ content. The largest diameter strange quark corresponds to part of the K^0 content. The middle diameter strange quark corresponds to part of the $K^\pm$ content. The smallest diameter strange quark contributes to the phi meson introduced in Section 15.6. The middle and smallest diameter up and down quarks correspond to the content of the pions (not shown)

Quarks and the structure of hadrons

Table 15.2 The set of quarks at 940 MeV

Quark	Charge Units of e	x_0		Thickness
Up	2/3	2π	6.28	$\pi/2$
	2/3	$5\pi/2$	7.85	"
	2/3	3π	9.42	"
Down	-1/3	2π	6.28	"
	-1/3	$5\pi/2$	7.85	"
	-1/3	$7\pi/2$	11.00	"
				"
Strange	-1/3	$\pi/2$	1.57[1]	"
	-1/3	π	3.14	"
	-1/2[2]	$3\pi/2$	4.71	"

Required for the phi meson, see Section 15.6
The strange quark at $x = 3\pi/2$ has a charge around $-0.5e$

We have given them names. Because some quarks contribute to more than one particle for which there are slightly differing values of the charge fraction, we have rounded the charge fractions to the values shown. The quarks shown in this table are the result of applying two principles. The first is that a quark at 940 MeV can transform into a constituent quark shown in Table 15.1 if the x_0 of the constituent quark lies within the x range (i.e. thickness) of the 940 MeV quark. The second principle is that only quarks that are in stable particles or long lived products can be generated in higher energy collisions. This is discussed further in Appendix H. The intention is that the quarks listed in Table 15.2 are necessary and sufficient to generate the quarks shown in Table 15.1.

This means that we can refer to the quark content of the particles from Table 15.1 as shown in Table 15.3 using the connections to the set of quarks in Table 15.2 (denoted by u, d and s) and their anti-particles

15.2 Quarks and the Dirac particle content of nucleons, kaons and pions

(denoted by $\bar{u}$, $\bar{d}$ and $\bar{s}$). Column 3 of Table 15.3 refers to the 940 MeV content enroute to forming the cylindrical forms of the kaons and pions, and the excited states of the proton and neutron, i.e. the composite forms indicated in the second column of Table 15.3, whereas in the standard model the various particles have the quark content shown in column four. This list of the contents of the various particles largely agrees with the standard model and hence the new theory predicts a major plank of the standard model and its account of hadron physics.

Table 15.3 The quark content of the particles from Table 15.1

Particle	Composite form	New theory 940 MeV x_{0i} and quark content en route to composite forms	Standard model quark content
p $\bar{p}$	Excited states	7.85/9.42/11.00 uud 7.85/9.42/11.00 $\bar{u}\bar{u}\bar{d}$	uud $\bar{u}\bar{u}\bar{d}$
n $\bar{n}$	Excited states	7.85/9.42/11.00 dud 7.85/9.42/11.00 $\bar{d}\bar{u}\bar{d}$	udd $\bar{u}\bar{d}\bar{d}$
π^+ π^-	Precursors	6.28/7.85 $u\bar{d}$ or $\bar{d}u$ 6.28/7.85 $\bar{u}d$ or $d\bar{u}$	$u\bar{d}$ $d\bar{u}$
π^0 $\bar{\pi}^0$	Precursors	Combination of $u\bar{u}$ and $d\bar{d}$ at 6.28/7.85	$(u\bar{u} - d\bar{d})/\sqrt{2}$ $(\bar{u}u - \bar{d}d)/\sqrt{2}$
K^+ K^-	Precursors	3.14/6.28 $\bar{s}u$ 3.14/6.28 $s\bar{u}$	$u\bar{s}$ $\bar{u}s$
K^0 $\bar{K}^0$	Composite particles	4.71/7.85 combination of $\bar{s}d$ and $\bar{s}u$ 4.71/7.85 combination of $s\bar{d}$ and su	$d\bar{s}$ $\bar{d}s$

Quarks and the structure of hadrons

Hence there is agreement with the standard model except for the neutral kaons where the new theory requires that there is a charge of $\mp e/2$ on one of the strange quarks and its anti-particle. The phi meson model based on the strange quark at $x_0 = \pi/2$ is discussed below in Section 15.6 and leads to a predicted particle energy of 1050 MeV and so strictly the third strange quark should not be included in the 940 MeV set, but it is convenient to include it in Table 15.2 at this stage in order to make the following point. There is a fundamental difference between the approaches of the standard model and the new theory. The standard model distinguishes between three versions of each quark flavour via the concept of colour. The new theory has three versions of each quark flavour due to three different x_0 values.

15.3 Gluons

The neutral parton has already been introduced with the excited states of nucleons in Chapter 14, Section 14.4, and is an independent oscillatory waveform composed of gravitational mass density with zero charge density. We are going to extend this concept of omega waveforms with oscillatory and steady state gravitational mass and call them gluons. They are produced by the mixing of oscillatory waveforms of quarks at different angular frequencies. The mixing process is discussed in Chapter 3 Section 3.4. Also gluons at different angular frequencies can mix producing further gluons. These reproduce some of the properties of gluons in the standard model. However, the new theory provides a physical explanation for them, whereas the standard model assumes that they exist and that they are the quanta of the strong interaction (Cottingham and Greenwood p2 (2007)). Gluons can be present under circumstances where the standard model invokes the strong force, but not all circumstances where the standard model invokes the strong force require gluons to be present in the new theory. For example the analysis of the nuclear interaction in Chapter 10 does not appeal to gluons.

Gluons are generated when there is overlap between shells at differing angular frequencies and again are independent oscillatory waveforms composed of gravitational mass density with zero charge density. The cases arising in the discussions which follow are

(1) The oscillatory components involved in the separation of quark and anti-quark in e^+e^- collisions, see Appendix H.

15.4 Quark production in e^+e^- collisions at around 940 MeV and above

(2) The oscillatory components due to mixing in $q\bar{q}$ particles ($s\bar{s}$, $c\bar{c}$, $b\bar{b}$), Section 15.6 et seq

15.4 Quark production in e^+e^- collisions at around 940 MeV and above

A number of facilities have contributed to the observation of the products of e^+e^- collisions. Lists can be found in text books; for example Krane (1988, p 597) gives details on the early facilities up to 1987, including the various experiments at the Stanford Linear Accelerator Centre. These were followed by the Large Electron – Positron collider (LEP) which was built in the circular tunnel at CERN. It achieved an energy of 209 GeV, the highest energy for any e^+e^- collider (Particle Data Group 2014). It was closed down in 2000, and replaced by the Large Hadron Collider.

With most e^+e^- colliders the centre of mass system is the laboratory system and so the total energy is shared equally between the two beams. Figure 15.2 shows a sketch of the collider configuration. Electrons and positrons enter from opposite directions along the centre line of the beam pipe. Jets of particles (hadrons) are emitted at approximately 90° to the beams, one each side of the beams.

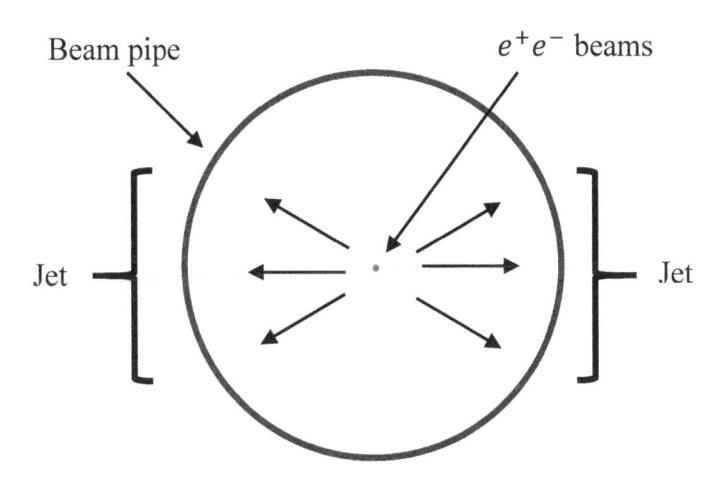

Figure 15.2 Formation of jets following an e^+e^- collision

Sometimes there are three jets emitted. In the standard model the

Quarks and the structure of hadrons

presence of a gluon is invoked to explain a three jet event but this is not investigated here. With the standard model, the conventional interpretation is that an initial $q\bar{q}$ pair is formed. These cannot separate and escape, and more $q\bar{q}$ pairs are created from the energy available and they form the jets of hadrons.

In the new theory we adopt the same approach that an initial $q\bar{q}$ pair is formed. We examine the consequences in more detail in Appendix H. From this examination it is concluded that there is conservation of quarks from the creation of the initial $q\bar{q}$ pair through to the particles in the jets, i.e. all the quarks created are contained within the particles in the jets. We also conclude that since the output particles can only contain quarks which contribute to the constituents of lower energy particles and therefore are in the 940 MeV set, then the first quark pair formed by the e^+e^- collision (the initial quarks) must be in the 940 MeV set. This conclusion is important because it allows the identification of the quarks which contribute to the total scattering cross section discussed below in Section 15.10. So there can only be the up, down and strange quarks. Hence the $q\bar{q}$ at 940 MeV are confined to $3u$, $3d$ and $3s$ (the extra s quark is discussed in Section 15.6).

A key feature of the 940 MeV set of quarks is that at 940 MeV they occupy filled and non-overlapping shells. Other than the one gap at $x = 9.42$ in the down quark sequence, they are continuous from $x = 0.785$ to 10.21, plus the down quark centred on $x = 11.00$. Of course these quarks are not confined to this energy. A single line through the origin on the x – energy diagram can represent one of these quarks resulting from an e^+e^- collision where the energy is up to that of the incoming e^+ or e^- in the centre of mass frame. So the quarks shown on Figure 15.1 are the quarks generated in e^+e^- collisions. We extend the list of quarks below but first a number of other topics need to be investigated.

15.5 The volume lines on the x – energy diagrams

Particles (mesons, baryons) formed at higher energies than 940 MeV can be composed of quarks from the 940 MeV set, ensuring that when they decay, they form stable or long lived particles of lower energy. We do not tackle the detail involved in deducing the structure of these

15.5 The volume lines on the x – energy diagrams

extra particles, but we can identify where they can be found on the x – energy plot by calculating the minimum and approximate maximum values of x for the particle outer boundary. First consider a spherical particle with radius x_s. The volume is given by

$$\frac{4\pi}{3} x_s^3 \left(\frac{c^3}{\omega^3}\right)$$

and in order for the particle to be in the subset, we equate this to (see Appendix D)

$$\frac{\kappa V_P}{K_P}\left(\frac{\omega}{\omega_{0P}}\right)^{1/2}$$

(15.1)

Putting $\kappa = 1$,

$$x_s = \left(\frac{3V_P}{4\pi K_P}\right)^{1/3} \left(\frac{\omega_{0P}}{c}\right)\left(\frac{\omega}{\omega_{0P}}\right)^{7/6}$$

Putting $E = \hbar\omega$ we can plot this expression on an x - energy plot as shown in Figure 15.3.

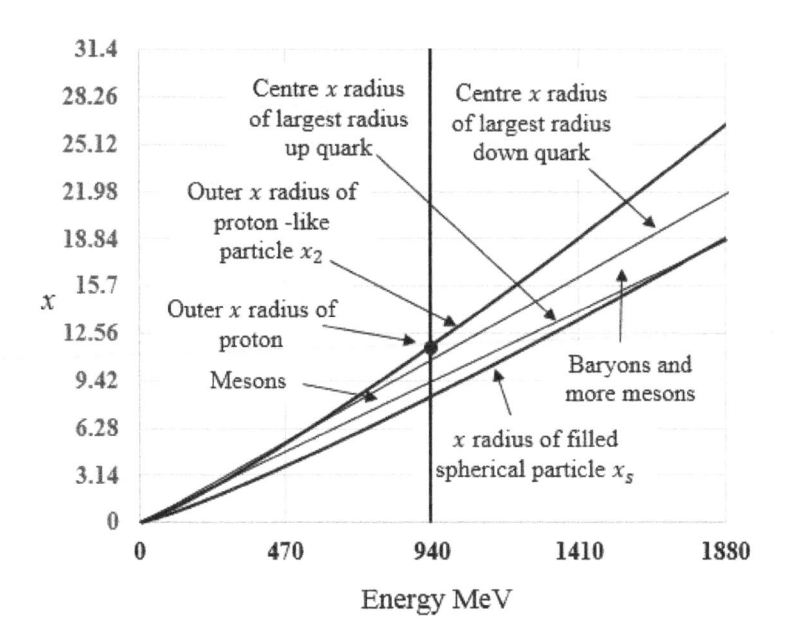

Quarks and the structure of hadrons

Figure 15.3 The x radius of a filled spherical particle, x_s, and the outer radius of a proton – like particle, x_2, versus energy. The x values for the outer boundaries of subset particles are confined to the regions between the x_s and x_2 lines. Because it is easier to fit in two shell particles rather than three shell particles, mesons occur in the region below 940 MeV and baryons and more mesons occur in the region above 940 MeV

Any deviation from a sphere will increase some parameter of the structure above x_s and in this sense x_s is the minimum value of x.

Next we consider a proton – structure type particle in which we reduce or increase x_2 (i.e. the outer cylindrical radius in x units) and make the volume equal to

$$\frac{x_2^3 \omega_{0P}^3}{x_{2P}^3 \omega^3} V_P$$

Equating this to (15.1) and putting $\kappa = \kappa_P$, the proton-like x_2 is given by

$$x_2 = x_{2P} \left(\frac{\omega}{\omega_{0P}}\right)^{7/6}$$

and this is also plotted on Figure 15.3. This is indicative of the maximum x extent in a structure.

The gap below the x_2 line and above the x_s line is where the x values of the outer boundaries of subset particles are confined. Kaons and pions are consistent with this picture, and the volume lines construction allows comment on the existence of more particles. The shrinking volume below the proton energy confines the extra particles to the two component mesons. The larger volumes available above the proton energy allow for more baryons and mesons, particularly those with extra angular momentum and therefore with large volume orbitals. Thus the picture described by the volume lines accords with what is observed experimentally. From the previous work in Chapters 2, 6, 9 and 14, there are many ways to construct these particles, some as single omega solutions, some as composite particles and some as composites with internal angular momentum. They can have cylindrical or spherical form. A key step is to predict their inertial masses. This is a large exercise which has not yet been attempted.

15.6 The phi meson

15.6 The phi meson

Having introduced the volume lines which constrain the position of subset particles on the x – energy diagram, we now introduce a particle, the phi meson, which is situated below the x_s line. This is on account of it not being in the subset as we shall see. In Table 15.2 we introduce a further quark to complete the 940 MeV set. The additional strange quark is shown in Figure 15.1 and at $x = \pi/2$ it is below the other strange quarks at $x = \pi$ and $3\pi/2$. We are unable to predict the charge fraction but because it occupies the lowest possible shell at 940 GeV we shall take it to have the lowest charge fraction of the other quarks i.e. 1/3. In order for this quark to be included it has to be demonstrated that a reasonably long lived particle can incorporate it. It is shown here that there is a particle, at 1050 MeV, and so although it completes the 940 MeV set in terms of x value, and its rest inertial mass is below 940 MeV/c^2, it will not be observed as an output quark until sufficient energy is available to create the particle at 1050 MeV. There may be other particles containing it at energies below 940 MeV, but these have not been investigated. In conventional treatments the phi meson is formed predominantly from the strange quark and its anti-particle (Halzen and Martin p 49 1984). Here we propose a structure containing the additional strange quark and its anti-particle.

The quarks cannot leave the composite particle, and so we have the s and $\bar{s}$ quarks shuttling from one end of the composite particle to the other, see Figure 15.4.

When there is complete overlap, then the analysis of Chapter 4 does not apply for reasons explained in Appendix J. This means that the formula for $\hbar$, equation (4.6), does not apply and the equations for the subset in Appendix D, Section D.5, in particular D.9 for the volume, do not apply. Thus the constraints described by the volume lines do not apply and the phi meson can have an x value below the x_s line on Figure 15.3.

Quarks and the structure of hadrons

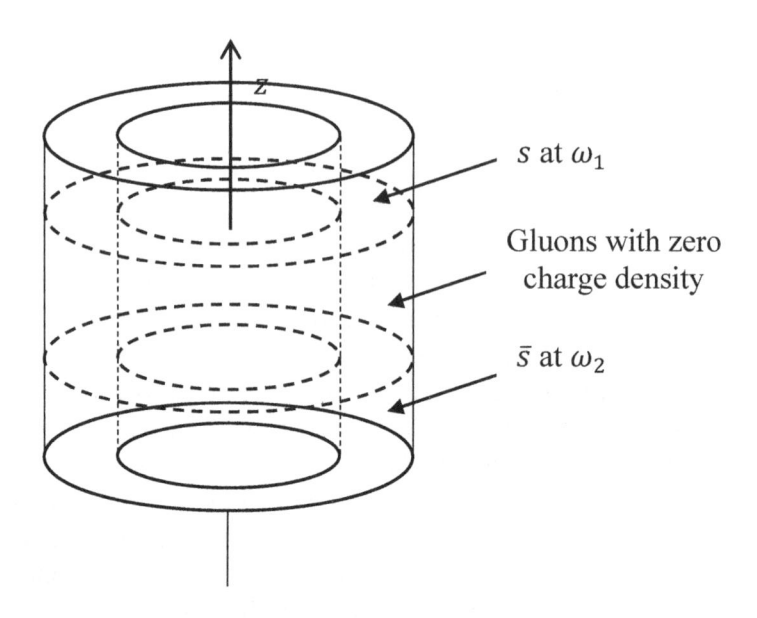

Figure 15.4 The proposed structure for the phi meson with overlap between the strange quark at $x = \pi/2$ at 940 MeV and its anti-particle. Oscillatory motion is along z through complete overlap followed by return in the reverse direction. Gluons are formed in the overlap region

Our model for this particle is as follows. The rest inertial mass for the 940 MeV lowest x_0 s quark is 598 MeV/c^2, see Figure 15.5. Each quark is extended along z, with additional motion normal to z. Interference with positive and negative charged shells leads to zero overlap potentials, see Sections 10A.4 and 10A.7 in the Book. By reducing the energy of either q or $\bar{q}$, they mix in the overlap region, with a gravitational mass oscillatory component. In the overlap region there is zero steady state charge density and so there is no electric charge density mixing component. This is consistent with the model for a gluon in Section 15.3.

15.7 The charm quarks and charmonium

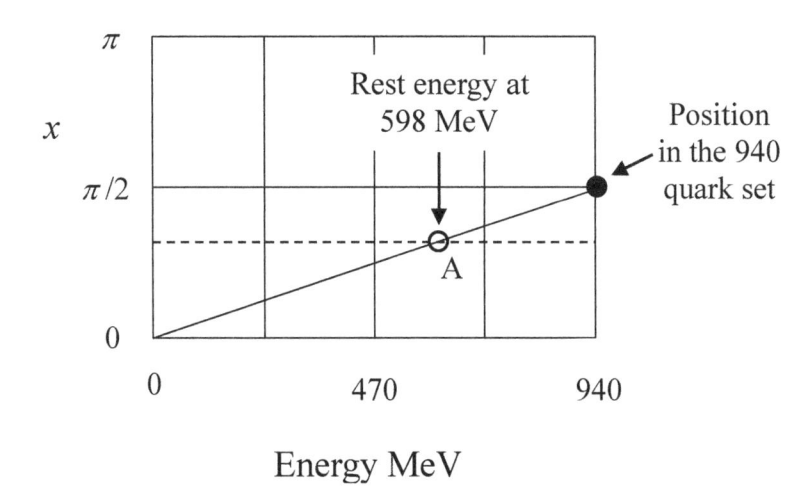

Figure 15.5 The extra strange quark at $x = \pi/2$ in addition to the other strange quarks at $x = \pi$ and $x = 3\pi/2$ in the 940 MeV set

The quarks cannot escape, and if there is motion of one quark there must be an effective binding energy. Let the angular frequencies of the quarks be ω_1 and ω_2. Hold ω_2 at ω_0 where $\hbar\omega_0/c^2$ is the rest inertial mass of each quark. The difference angular frequency $\omega_2 - \omega_1$ is the gluon frequency. The gluon energy is $\hbar(\omega_2 - \omega_1)/2$. So the total energy is $\hbar(3\omega_2/2 + \omega_1/2)$. When $\omega_1 = \omega_2 = \omega_0$ the energy is $2\hbar\omega_0$. When $\omega_1 = 0$ the energy is $3\hbar\omega_0/2$. Averaging the energy over all values of ω_1, the energy becomes $7\hbar\omega_0/4$. The same result is obtained by interchanging ω_1 and ω_2. Using the rest energy of the lowest x_0 s quark of 598 MeV, the predicted rest energy of the phi meson is 1050 MeV compared to the measured value of 1019 MeV (Particle Data Group 2014).

15.7 The charm quarks and charmonium

We now introduce the charm quark indicated by c. Start with charm quarks placed at $x = \pi, 2\pi, 3\pi$ at 1880 MeV and each spans two shells as shown in Figure 15.6. Thus there is consistency (ie no overlap) with the 940 MeV set extrapolated to 1880 MeV. The justification for this scheme is that particles $c\bar{d}$ and $c\bar{u}$ can be formed at 1880 MeV which

Quarks and the structure of hadrons

satisfy the volume constraint as illustrated in Figures 15.7 and 15.8. 1880 MeV is close to the observed energies for $c\bar{d}$, 1870 MeV, and for $c\bar{u}$, 1865 MeV (Particle Data Group 2014). These are relatively long lived particles (the lifetime is of the order of 10^{-12} s) and provide a route through which charm quarks have a connection with lower energy particles.

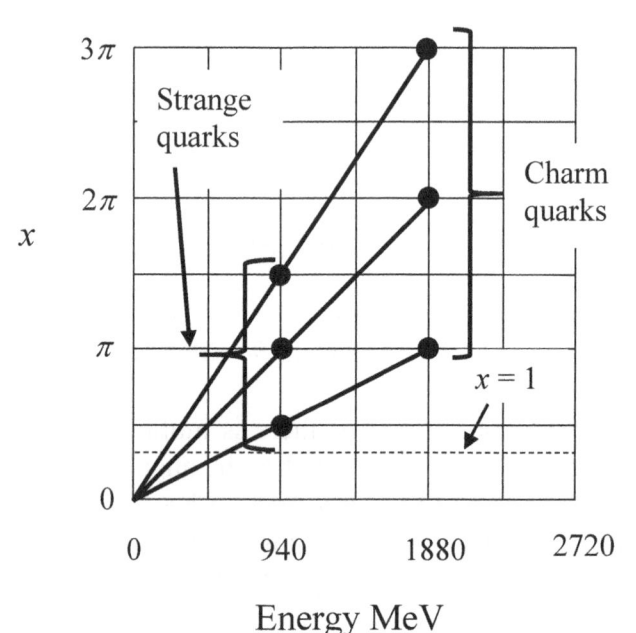

Figure 15.6 The charm quarks introduced at 1880 MeV

Consider the neutral $c\bar{u}$ first, Figure 15.7. Construct a single omega solution mixture of the three charm quarks so that the charge of a single quark is spread over the three shells shown on Figure 15.7. A similar single omega solution is formed by a mixture of the three up quarks at a lower angular frequency, shown by the connected squares at 1880 MeV. So the set of three charm quarks are combined in a superposition providing $2e/3$ charge and similarly the up quarks provide $-2e/3$ charge. A composite particle is formed by the combination of these cylindrical single omega solutions and this is the precursor to a spherical single omega solution particle with spin zero (Group A, Chapter 2, Section 2.4). The composite particle has an energy of 1880 MeV because this is the energy at which the cylindrical

15.7 The charm quarks and charmonium

quark shells cover exactly two spherical shells each, indicated on Figure 15.7.

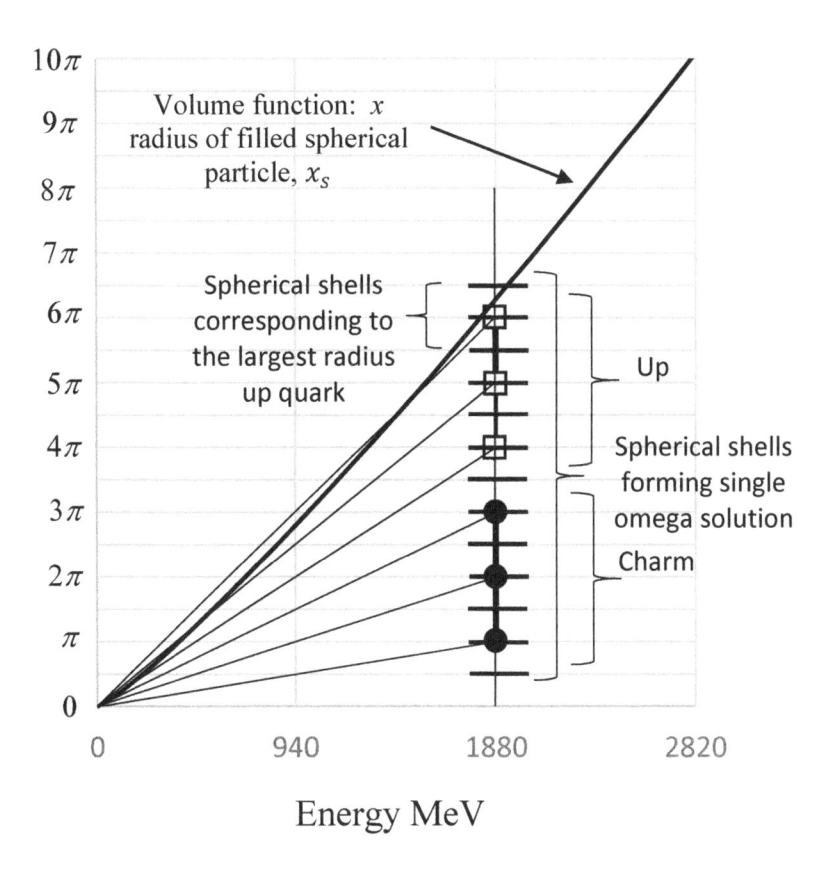

Figure 15.7 The $c\bar{u}$ meson. The spherical single omega solution is composed of the filled spherical shells, shown separated by short horizontal lines, extending up to the x_s line. The composite precursor particle is based on the two cylindrical single omega solutions, one provided by the charm quarks (filled circles) and the other by the up quarks (squares). Each quark at 1880 MeV extends over two spherical shells

Most of the composite particle energy is provided by the charm quarks. This energy is above the charm rest mass energy due to internal motion, for example due to extra angular momentum and spreading of the densities with r_c as well as motion along z. However the composite particle energy will be increased by the contribution from the up quarks and reduced by any binding energy due to overlap of the charm

Quarks and the structure of hadrons

and up quarks, and so a simple estimate of the charm energy is 1880 MeV.

Although there is a gap at the lowest x, the spherical single solution particle at energy 1880 MeV has a small extension above the x_s line. A similar analysis to that above for the $c\bar{u}$ meson can be applied to the $c\bar{d}$ meson, depicted in Figure 15.8.

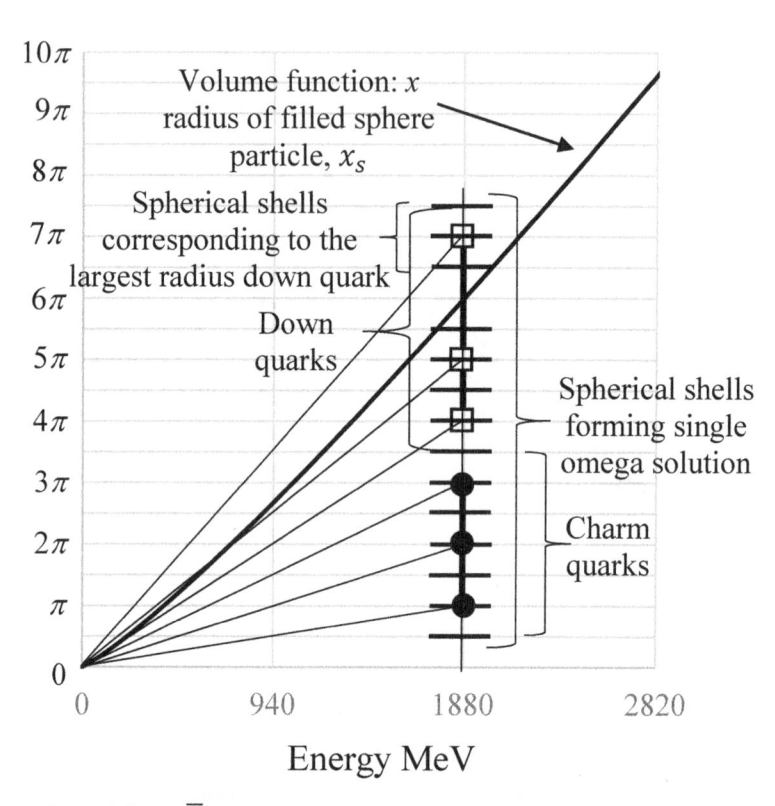

Figure 15.8 The $c\bar{d}$ meson. The spherical single omega solution is composed of the filled spherical shells, shown separated by short horizontal lines, extending up to the x_s line. Note missing spherical shells between $x = 13\pi/2$ and $x = 11\pi/2$. The composite precursor particle is based on the two cylindrical single omega solutions, one provided by the charm quarks (filled circles) and the other by the down quarks (squares). Each quark at 1880 MeV extends over two spherical shells

A similar conclusion regarding the volume, that the volume is that

15.8 The bottom quark

implied by the x_s line, applies to the $c\bar{d}$ meson where there is a further gap at at $x = 6\pi$ but there is a compensating extension above the x_s line.

The standard model takes $c\bar{c}$ as being its model for the charmonium particle. Here we also propose a structure containing c and $\bar{c}$. Using 1880 MeV as the charm quark energy, we can calculate the charmonium energy in the same way as for the phi meson, giving (7/4)1880 = 3290 MeV for the charmonium rest energy. The observed values are 2984 MeV for the $\eta_c(1S)$ state and 3097 MeV for the $J/\psi(1S)$ state (Particle Data Group 2014) and this approximate agreement provides further evidence that the charm quarks have been placed at about the right energy. The details in this section are qualitative and require rigorous quantification.

15.8 The bottom quark

Can the new theory account for the bottom and top quarks of the standard model? Maybe both cases can be included in the new theory. Here we propose the origin of the bottom quark and attempt to predict the rest energy of bottomonium, which in the standard model is taken to contain $b\bar{b}$. Let's start by using the observed energy of bottomonium, 9460 MeV (particle Data Group 2014), to calculate the quark rest energy required using the phi meson model. In this case the rest energy required is 9460/1.75 = 5400 MeV and the quark energy when occupying a shell centred on $x_0 = \pi/2$ is 8460 MeV, shown on Figure 15.9. The lower edge of the nearest charm quarks at 1880 MeV is shown, passing through $\pi/2$ at 1880 MeV, resulting in $x' = x - \pi/4 = 0$ at 940 MeV, and the bottom quarks have to fit in below this line. At 8490 MeV it is possible to fit in four quarks. The lowest energy at which it is possible to fit in three quarks is 6580 MeV and the lowest x_0 quark has a rest energy of 4189 MeV corresponding to a bottomonium energy of 7330 MeV using the 1.75 factor, which is significantly less than the measured value. However the various decay paths for bottomonium have not been examined and there may be a requirement for a minimum energy above this value in order for decays via other established quarks to be possible. So the number of bottom quarks has not been determined and more work is required.

Quarks and the structure of hadrons

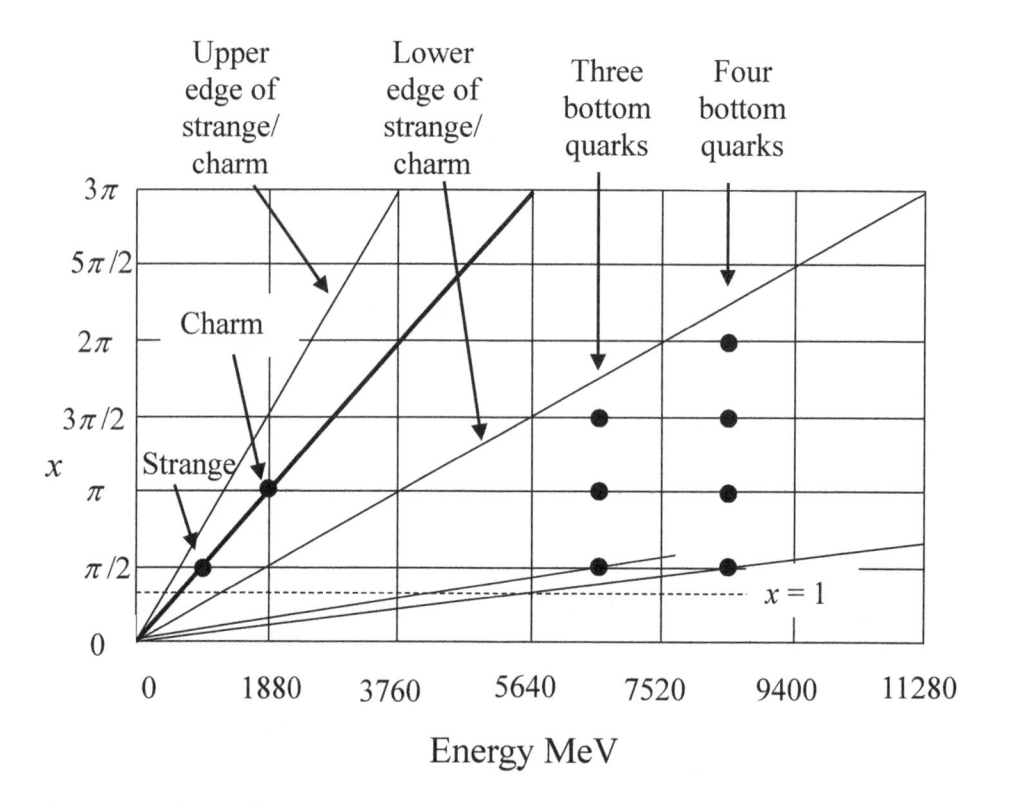

Figure 15.9 Bottom quarks. The lowest strange and charm quarks are shown. The bottom quarks have to fit in below the lower edge of the strange/charm quarks. Three bottom quarks are possible at 6580 MeV and four are possible at 8490 MeV

15.9 Summary of particle predictions

Table 15.4 collects together the results from this chapter and Chapter 14, and extends the results shown in Table 8.1. Appendix J collects together the various categories of particle.

15.10 The scattering ratio *R*

Table 15.4 Particle properties

Particle	Form	Structure	Predicted inertial mass × c^2 MeV	Measured inertial mass × c^2 [1] MeV
$K^\pm$	Spherical with cylindrical precursor	$\bar{s}u, s\bar{u}$	507	493.7
K^0	Cylindrical and spherical forms	Combination of $\bar{s}d$ and $\bar{s}u$	513	497.6
Phi	Overlap	$s\bar{s}$	1050	1019
$c\bar{u}$	Spherical single omega solution with $c\bar{u}$ precursor	$c\bar{u}$	1880	1865
$c\bar{d}$	Spherical single omega solution with $c\bar{d}$ precursor	$c\bar{d}$	1880	1870
Charmonium	Overlap	$c\bar{c}$	3290	2984 $\eta_c(1S)$ 3097 $J/\psi(1S)$

Particle Data Group (2014)

15.10 The scattering ratio *R*

In Chapter 13 we have established that quantum field theory and quantum electrodynamics follow from our theory. Hence the conventional treatment of $e^+e^- \to \mu^+\mu^-$ and the extension to quark production $e^+e^- \to q\bar{q}$ also follows, and which include the Feynman diagrams in Figure 15.10 (a) and (b). The $q\bar{q}$ production diagram is used in Appendix H. The question is, what is the set of quarks involved? Our attention is

Quarks and the structure of hadrons

directed to the situation at an energy at around 940 MeV and above.

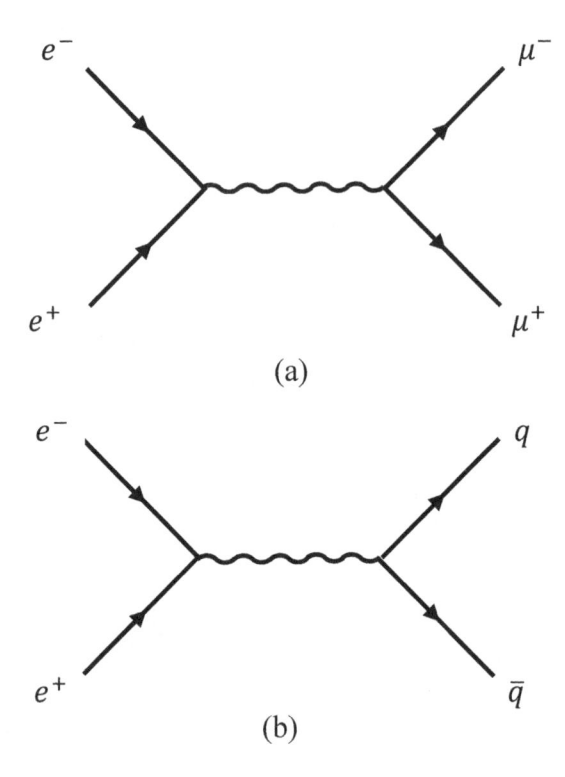

Figure 15.10 Feynman diagrams (a) $e^+e^- \rightarrow \mu^+\mu^-$ (b) $e^+e^- \rightarrow q\bar{q}$

We have demonstrated in the preceding sections above that at 940 MeV:
(1) Quark charge magnitudes should be either $1/3e$ or $2/3e$. It is not clear whether the fractions are exact or whether there are small variations about these values.
(2) There are nine quarks involved at 940 MeV, three down, three up and three strange as shown in Figure 15.1 and this agrees with the standard model.

The particles involved in Figures 15.10 (a) and (b), electrons, muons, quarks and their anti-particles, all have either a ring structure or cylinder structure. All have a central radius such that when $c/\omega r_c = 1$, they have the Dirac magnetic moment and share the same spin

15.11 The connection between the new theory and the Standard Model

functions with a spin angular momentum of $\hbar/2$. Since quantum field theory and electrodynamics follow from the new theory, then we can use the standard formula for the production cross-section of a charged Dirac particle and its anti-particle from an e^+e^- collision. The cross section is proportional to the square of the output particle charge. This means that we can introduce the ratio R where

$$R = \frac{\sigma(e^+e^- \to \text{hadrons})}{\sigma(e^+e^- \to \mu^+\mu^-)} = \sum_i c_i^2$$

where the σ's are cross sections, the c_i are the charge fractions and the sum is over all quarks. At 940 MeV R becomes 2 both in the standard model and the new theory. Inclusion of the charm quarks at 1880 MeV raises R to 10/3 both in the new theory and in the standard model. There are three bottom quarks in the standard model and R rises to 11/3. The agreement with experiment is good above 11 GeV (see for example Cottingham and Greenwood p 14 (2007)). Since it is not yet clear how many bottom quarks are predicted, the new theory cannot provide a value for R at the higher energies.

15.11 The connection between the new theory and the Standard Model

We summarise the aspects where the new theory is in agreement with the standard model's account of the structures and physics of hadrons, and where the new theory is in disagreement. In so doing, ideally, we require an incisive critical review of where the standard model works and where it does not, and, in order to continue the success of the new theory, that the new theory should explain why the standard model is successful and where it is not successful. However this review is not to hand, and if it were this appraisal may need revision. In what follows we list the similarities and differences under various headings.

Composition of Hadrons. The list of quarks in Table 15.2 plus the charm and bottom quarks agrees with the standard model. However it is not yet clear in the new theory how many bottom quarks there are, and the new theory cannot yet account for the top quark. It is a success of the standard model to have deduced the system of quarks from experimental data. However, treating the standard model as a formal

Quarks and the structure of hadrons

system, the quarks are assumed, whereas we have derived their existence and their properties from the two postulates. Thus there is considerable agreement, as evidenced by Table 15.3, with the standard model. However the new theory has not attempted to examine the structure of the vast array of known other mesons and baryons. In the standard model the description is confined to reference to what we call the 940 MeV set together with charm, bottom and top quarks, whereas our account of structures distinguishes between compositions labelled using the 940 MeV set, the actual quark content of composite particles, and single omega solution particles.

Colour. The standard model distinguishes three types of colour (red, green and blue) for quarks of the same flavour (up, down, strange, charm, bottom and top). This is not a concept in the new theory where instead the quarks of each flavour (up, down, etc.) have three different values of x_0. So the new theory does not lead to quantum chromodynamics. In the standard model the quark rest inertial mass are the same for all colours of a particular flavour. In the new theory the rest inertial masses are different and so no attempt has been made to compare predicted rest inertial masses with experimental results.

Binding, asymptotic freedom. In the standard model quarks are stated not to be able to exist outside bound structures, but inside such structures they have freedom of movement. There is agreement with the standard model, but the new theory gives an explanation. The quark fractions of e charges means that they cannot exist outside composite particles, because they cannot be accelerated by existing photons. The new theory's account of photons in Chapter 3, Section 3.5 leads to independent particle charges being confined to $\pm ne$ where n is an integer including zero.

Gluons. Qualitatively the account of gluons in Section 15.3 has similarities with those of the standard model, but the new theory gives a mechanism (mixing) for the generation of gluons, and gives a description (Type A waveforms) for them. The neutral parton has already been introduced with the excited states of nucleons in Chapter 14 and is an independent oscillatory waveform composed of gravitational mass density with zero charge density. Gluons are generated when there is overlap between shells at differing angular frequencies and again are independent oscillatory waveforms composed of gravitational mass density with zero charge density. The cases arising in the discussions above are:

15.11 The connection between the new theory and the Standard Model

(1) The oscillatory components involved in the separation of quark and anti-quark in e^+e^- collisions
(2) The oscillatory components due to mixing in $q\bar{q}$ particles ($s\bar{s}$, $c\bar{c}$, $b\bar{b}$)

There is a further case which has not been considered:

(3) The movement of quarks with radial or angular momentum in composite particles resulting in overlap and mixing

R analysis. The new theory has not investigated the top quark and neither can it say how many bottom quarks there are. However at lower energies its account of the ratio R in Section 15.10 is identical with that of the standard model.

Symmetries. Symmetry has already been discussed in the context of the new theory and QFT in Section 13.4. Maybe the various particle type conservation laws (lepton number, baryon number, meson number) can be deduced from the principle of gravitational mass and charge configuration matching. As noted above, gluons in the two approaches are similar and so maybe we can accept into the new theory the Feynman diagrams containing quark interactions involving gluons. In the standard model colour quarks are in all other respects identical. In the new theory the parameter corresponding to colour is the x_0 value, and these differ significantly from each other. So treating these quarks as identical under a symmetry operation is a simplification.

Conclusions. We conclude that there are areas where the new theory leads to the standard model analysis (flavours of quarks, content of particles, existence of gluons). The standard model account of hadronic physics based on assuming these concepts follows from the new theory by deduction from first principles. There are areas where there is a possible difference in detail (e.g., number of bottom quarks) and there are areas where there is disagreement (colour in the standard model versus x_0 in the new theory). Refinement of this assessment requires more work. There is much more detail to be attended to in the development of the new theory in the following areas:

(1) The structure and properties of the mesons and baryons between the two volume lines
(2) The existence or otherwise of the top quark.
(3) Does the theory predict more particles at higher energies?
(4) The investigation of decays and particle interactions

Quarks and the structure of hadrons

15.12 Summary

This chapter has developed the account of quarks within the new theory. The introduction of a set of quarks at 940 MeV (up, down strange) is a bridge between what is generated in e^+e^- collisions and the independent Dirac particle content of composite forms of nucleons kaons and pions. Gluons are introduced as the product of mixing between quarks. Areas are identified on the x – energy plot where mesons and baryons can occur. The scheme of quarks is extended to include charm and bottom quarks. The model for the phi meson, in which a quark – anti-quark pair can form a particle with a volume lower than the subset volume, is applied also to charmonium and bottomomium. A comparison between the results for the new theory and the standard model highlights where there is agreement, and where there are differences.

Chapter 15 references

Cottingham W N and Greenwood D A 2007 *An introduction to the standard model* Second Edition Cambridge
Halzen F and Martin A D 1984 *Quarks and Leptons* Wiley
Krane S K 1988 *Introductory Nuclear Physics* Wiley
Particle Data Group, *Review of particle physics*, Chinese Physics C Vol 38, No. 9 (2014) 090001

Chapter 16

Weak processes

16.1 Introduction

The previous chapter shows the extent to which the new theory predicts features of the standard model concerning quarks and their role in hadron physics. There is another major part of the standard model which is concerned with weak interactions and decays. Its development leads to the introduction of the Higgs field, to an account of the origin of inertial mass and to the Higgs particle, see ,for example, Cottingham and Greenwood (2007 pp 105, 115). The detection of a new particle consistent with it being the Higgs particle was announced by CERN at a seminar on 4 July 2012 (Baggott 2012 pp215 – 219) and there are books describing the background to this discovery (Baggott, Carroll 2012).

It is difficult to untangle those results in these areas which are derived from the formalism of the standard model without further assumptions, and those areas which are dependent on new assumptions which should be declared as formal postulates. Without this formal structure being made explicit, it is difficult to decide which parts of the standard model in this area follow from the new theory. This examination of the standard model's formal structure is not tackled here. Nevertheless there are comments to be made.

As mentioned in Chapter 13, Section 13.4, QFT works exceedingly successfully because of the following features. Certain particle properties are assumed for the particles involved. Certain symmetries and/or conservation rules are assumed for the process. This means, from the point of view of the new theory, that at the particle level, the input and output particle mixes have the same, say, electric charge distribution, gravitational mass distribution, spin angular momentum and maybe other features, so that they are interchangeable, i.e. they have essentially the same physical description. The input and output differ in having different wave functions, from energy states obtained

Weak processes

as solutions of Schrodinger's equation or of Dirac's equation, and therefore different distributions of the particle mix. Perturbation theory is used to obtain the transition probabilities, transition rates and scattering cross sections. This involves the perturbation potential energy. The edifice of QFT then follows. Hence QFT depends on the three inputs: particle properties, symmetries/conservation rules and the potential energy expressions appearing in Hamiltonians and Lagrangians. So the new theory has to provide particle models, symmetries/conservation rules and the interaction potential energies, and then the QFT analysis of weak interactions will follow.

In addition within the standard model there is the inclusion of the Higgs mechanism from which it is claimed that the Higgs field is required to imbue particles with inertial mass. The new theory does not need this process since it already has an account of the origin of inertial mass. So this is where there is a parting of the ways. If the new theory does not require the Higgs mechanism, then how does it account for the CERN observation of a new particle at 125.7 GeV (Particle Data Group 2014)?

The various points above add to a large agenda. How far have we got with it? Not very far. In the sections below we restrict ourselves to comment on beta decay and we propose models for the W and Z particles.

16.2 Beta decay

Our starting point is the comparison of the proton and neutron structures determined in Chapter 9 and shown in Figure 16.1. It can be seen that shell 1 of the neutron can be described as the superposition of a shell of charge $-e$ and the corresponding proton shell with charge $+2e/3$. The relative extents of these entities along z shown in Figure 16.1 are only in order to see them separately. It is proposed that the $-e$ charge is lost from the neutron either in the form of an expanding charge ring as shown on Figure 16.2, or in the form of a sustained outward flow of negative charge. We consider the case where the electron and anti-electron-neutrino are distributed by radial wave functions. Consider a wave function which has the effect of reproducing the electron ring but at a smaller radius than that for the basic particle.

16.2 Beta decay

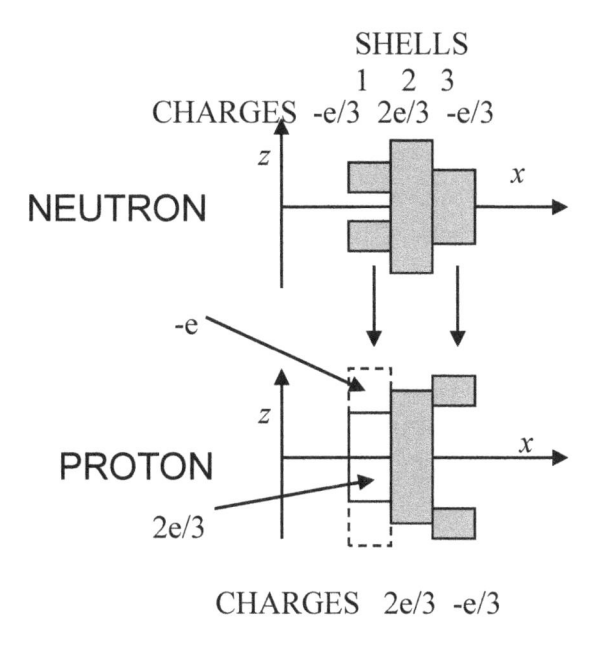

Figure 16.1 The structure changes involved in the decay of a neutron into a proton

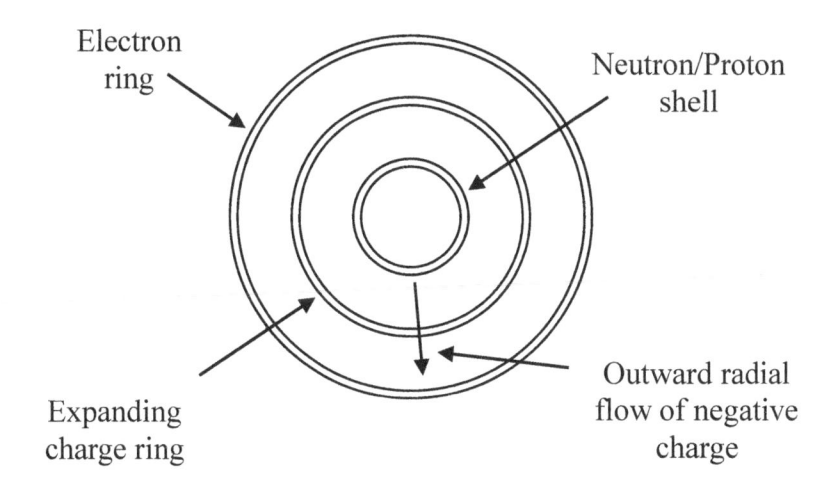

Figure 16.2 The charge flow outwards from the decaying neutron

Weak processes

The expanding charge ring can be described by the superposition of the effect of this wave function on the electron and its effect on the anti-neutrino. The sustained charge flow can be described by a further distribution of this superposition. When the charge ring reaches the electron radius, the electron and neutrino separate. The high angular frequency of the electron-anti-neutrino composite is manifest as kinetic energy of the two independent particles with their separate motions along z.

Note that in beta decay, the expanding ring of charge needs to have its own solution of the Field Equations, or is it the superposition of a neutrino and an electron which cannot separate until their free space radius is achieved, or does it involve a W particle? These issues have not been resolved, but a suggestion is made in the next section for models of W and Z particles.

16.3 W and Z particles

The existence of $W^{\pm}$ and Z particles are major predictions from electroweak theory, and they have measured inertial masses $\times c^2$ of 80.385 GeV and 91.1876 GeV respectively (Particle Data Group 2014). In Chapter 3 we show that there are travelling wave solutions of the Field Equations which couple together an oscillatory gravitational mass density waveform and an oscillatory electric charge density waveform. Each of these is composed of a sum of a number of components, labelled by n, each component consisting of a product of a Hermite function and a sinusoidal carrier. The Type A waveform is where the amplitude of the charge density components sum to zero in the central part of the waveform, and this is the case chosen to model $W^{\pm}$ and Z particles. Associated with the oscillatory gravitational mass density waveform is a steady state gravitational mass. We propose the following model for the $W^{\pm}$ and Z particles. In order to achieve a particle which makes use of the photon type waveforms of Chapter 3 and which is localised as a particle which can be stationary or have a velocity less than the speed of light, we propose that the radiation propagates round a circular path within the particle. We imagine these travelling waves to be bent into circular motion, one travelling in an outer ring shown in Figure 16.3, and a second one in an inner ring, so that we can have a number of large n waveforms around

16.3 W and Z particles

the circumference of the cylinder. With s wavelengths in ρ in the outer ring, $s-1$ wavelengths in the inner ring, where s is an integer, and $s-1$ wavelengths in m in the outer ring, $s-2$ wavelengths in the inner ring and with counter rotating rings the angular momentum can be $\hbar$. The volume of the particle is to be that required for membership of the subset. The Hermite function end sections give rise to two rotating charge spikes (either both positive or both negative) and this ensures that the steady state charge density, ρ_B, locally is greater than $1/A_\rho$. The charges in the two rings add for $W^\pm$ particles, and the charges cancel for Z particles.

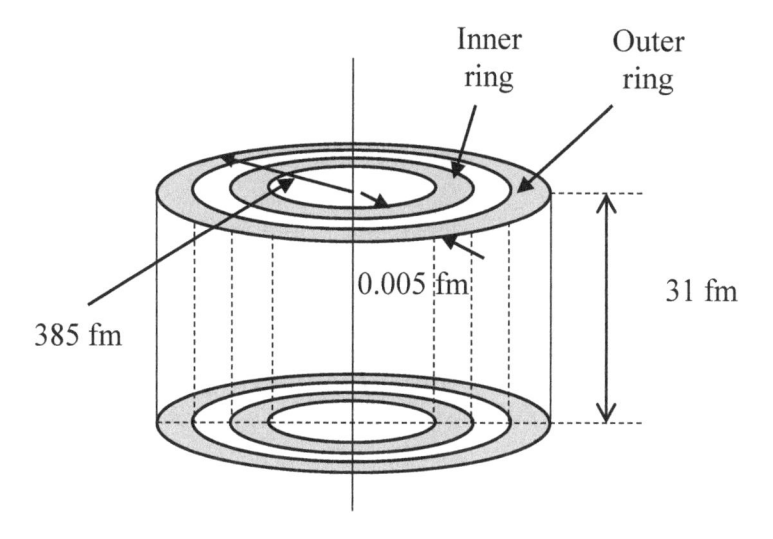

Figure 16.3 The proposed model for $W^\pm$ and Z particles. Each ring has a radial thickness of 0.0017 fm and are separated by 0.0017 fm

The model assumes that the radius of the cylinder is that of the electron ring. The analysis in Appendix I results in a predicted inertial mass of 119 GeV c^{-2} for both the $W^\pm$ and Z particles. Another way of expressing this conclusion is that, in order that the experimental energy values to be obtained, the cylinder radius needs to be around that of the electron. A consequence is that the $W^\pm$ particles can readily convert to a positron or electron plus the appropriate neutrino, but it does not progress the resolution of the issue as to how the new theory is to model beta decay discussed in the previous section.

Weak processes

16.4 Summary

This chapter tackles the topic of weak interactions and decays. The development of the standard model is very detailed in this area leading to electroweak theory and the Higgs mechanism. The extent to which the new theory predicts this detail is not yet clear. Two topics are tackled. An account is given of the mechanism of beta decay and which involves a ring of charge being emitted which results in emission of an electron and a neutrino. $W^{\pm}$ and Z particles are important components of the weak scene and models are proposed for them leading to a prediction of their inertial masses.

Chapter16 references

Baggott J 2012 *Higgs – the invention and discovery of the 'God Particle'* (Oxford)

Carroll S 2012 *The particle at the end of the universe* (Dutton Penguin Group)

Cottingham Q N and Greenwood D A 2007 *An introduction to the standard model of particle physics* (Cambridge)

Particle Data Group 2014 *Review of particle physics* Chinese Physics C Vol 38, No. 9 (2014) 090001

Chapter 17

The background

17.1 Introduction

In this Chapter we develop the concepts of the background and the near compensating boundary charge and gravitational mass in particles. The background provides a mechanism for the propagation of photons in what otherwise would be empty space.

17.2 The background

In Chapter 2, Section 2.8, we propose that free space is filled with a thinly distributed particle background. We develop this concept by proposing that the background particles, which need to be identified, are each distributed by a wave function ψ, such that $\psi\psi^*$ is uniform and the background steady state charge and gravitational mass densities are

$$\rho_B \int_V \psi\psi^* dV$$

$$m_B \int_V \psi\psi^* dV$$

where V is the basic particle volume and ρ_B and m_B are the particles' internal steady state densities. In other words, since the background particle is distributed over a volume V_D, then

$$\int_{V_D} \psi\psi^* dV = 1$$

and the particle internal densities are reduced by a factor

The background

$$\int_V \psi\psi^* dV = \frac{V}{V_D}$$

This means that the external potentials from source particles, ρ_{BE} and m_{BE}, are also reduced by the same factor, and that omega waveforms and photon waveforms travel in the distributed interiors of particles everywhere. A further consequence is that the steady state gravitational mass and charge densities reduce to exceedingly low levels at the external surface of particles and these density levels continue into the background.

This means that we require thin high density layers of charge and gravitational mass of opposite sign at the particles' external surface that nearly cancel the internal values. We call these the near compensation charges and gravitational masses. Figure 17.1 sketches these ideas.

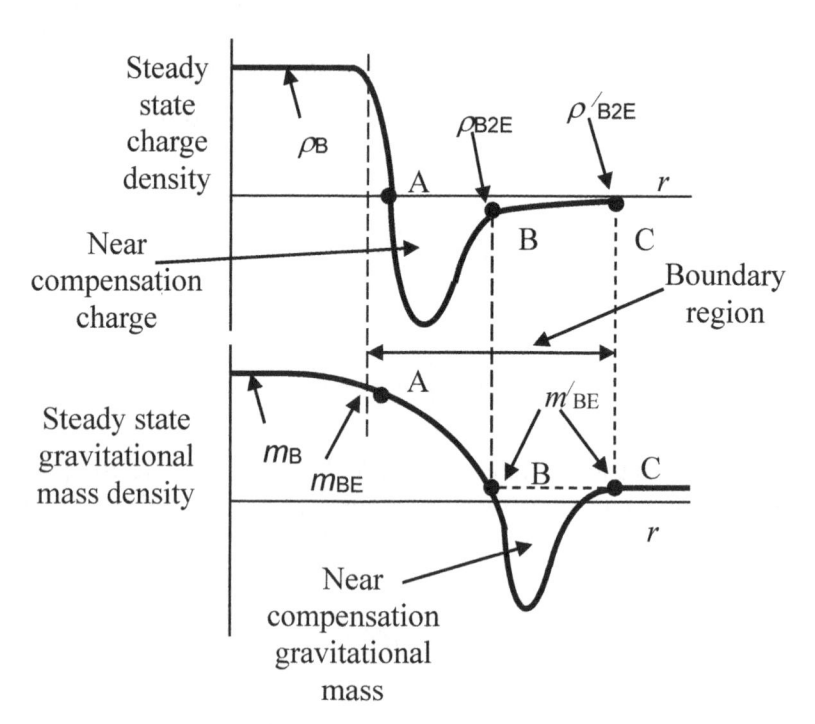

Figure 17.1 Proposed model for the particle boundary region

17.2 The background

The extent of the boundary region is exaggerated compared to the interior. m_{BE} the 'external' gravitational mass density due to the electric potential, becomes the value just inside the compensation charge shell at point A. The level at B and in the background beyond the compensation shells at C is

$$m'_{BE} = m_{BE} \int_V \psi\psi^* dV$$

ρ_{B2E} appears immediately after the near- compensation charge at point B and reduces to

$$\rho'_{B2E} = \rho_{B2E} \int_V \psi\psi^* dV$$

beyond the near-compensation gravitational mass at C. The net charge i.e. particle charge plus compensation charge, for a particle of charge e is

$$e \int_V \psi\psi^* dV$$

in order for Maxwell's equations to apply in the background. Similarly the net gravitational mass is

$$M \int_V \psi\psi^* dV$$

in order for the gravitational-gravnetic equations to apply in the background. For fields within basic particles, Maxwell's equations and the gravitational-gravnetic equations apply as though the potentials seen internally continue via m_{BE} and ρ_{BE} (and then ρ_{B2E}) into the intervening space. The model also means that all processes take place inside particles, whether within discrete particles or within the distributed particles in the background. In the Book we give two ways of estimating the reduction factor

The background

$$\int_V \psi\psi^* dV$$

leading to 10^{-37} and 2.3×10^{-39}. There is discussion in the Book examining the alternatives of neutrons, hydrogen atoms and electron neutrinos providing the background particles, with the neutrino being the most likely candidate. The transport by photons of gravitational mass and energy in the background is discussed in Chapter 12 in the Book and in Appendix B in this book.

17.3 Electrons and electron neutrinos

In the light of the model above for the structure of the boundary of particles, we can now offer an explanation for the electron having positive gravitational mass and negative charge. In Chapter 7 we temporarily said that although there is continuity in the oscillatory gravitational mass density at the boundary, there is a discontinuity in the steady state gravitational mass density to allow it to change sign. In Chapter 7 it is proposed that the electron is a hollow toroid with a thin shell. Surround the outer surface of the thin shell with another thin shell with a similar structure and similar magnitude of charge but with opposite sign of charge – this is the near-compensating charge discussed above. This is shown in Figure 17.2 (a) where r is now the radius indicated in Figure 17.2 (b). In order for the particle to have a net negative charge, the outer structure can be positive with a negative gravitational mass, surrounding a negative charge with positive gravitational mass. This structure ensures that when the charge is negative the gravitational mass is positive and vice versa, required by equation (7.2) in Chapter 7. Note that this proposal is a revision of the one presented in the Book, Chapter 12, Section 12.11.

We propose that the electron neutrino has the electric charge structure of Figure 17.2, but with the negative charge compensating the positive charge exactly. The positron is proposed to have a similar structure to the electron but with a net positive charge resulting from the outer positive charge being greater than the inner negative charge's magnitude. So we have now resolved the issue posed in Chapter 7 as to why the neutrino has a two ring structure and the electron and positron had only one – we now propose that all three have a two ring structure.

17.3 Electrons and electron neutrinos

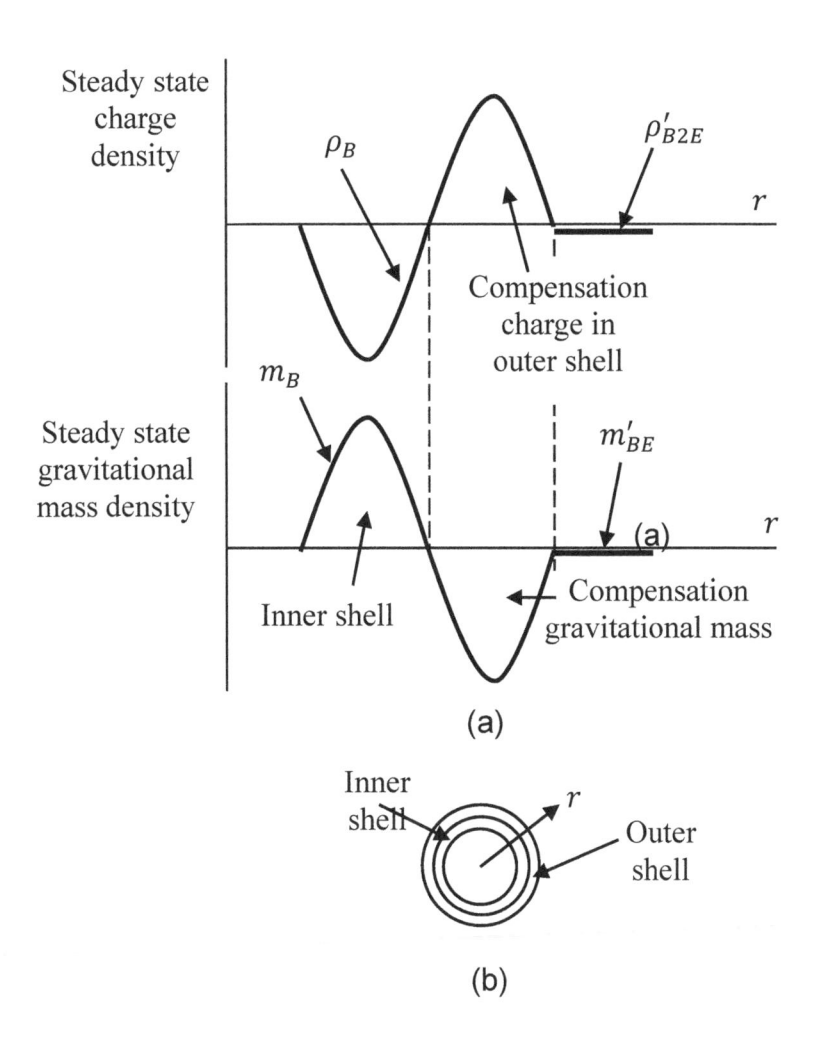

Figure 17.2 (a) Steady state charge and gravitational mass densities within the proposed structure of the electron. The original densities are in the inner shell and the compensation densities are in the outer shell (b) r is the radius from the centre of the circular cross section of the particle ring circumference. A further inner compensation shell is required to complete the structure

The background

17.4 Charge densities due to distant objects

In this section we examine the charge densities inside particles which are commandeered by gravitational potentials from distant objects. We commented above that large charge densities throughout the universe due to the permeation of gravitational potentials do not occur because of the reduction factor multiplying ρ_{B2E} in the background. However the ρ_{B2E} tails from distant objects, without a multiplying reduction factor, will be summed in discrete particles and we need to assess the magnitude of this effect. The external charge density from a gravitational mass M is $M/4\pi\epsilon_0 r$, where

$$M = \frac{Ac^2 \mathcal{M}}{\hbar} = 9.4 \times 10^{-11} \mathcal{M} \; C$$

where $\mathcal{M}$ is the inertial mass in kg.

Table 17.1 shows the resulting external charge density commandeered within a target particle for four inertial masses, to be compared with a proton internal charge density of 4×10^{24} Cm^{-3} (Appendix D, Figure D.7). The cases examined are at the surface of the sun, at the sun's position in the galaxy, and at our position at the centre of the surrounding universe for the whole mass of the universe and for that due to visible matter. This table is a corrected version of Table 12.3 in the Book. The conclusions are the same as previously. The results show that for the surface of the sun and at our location in the galaxy the levels are a small fraction of the internal proton charge density. However for the total known universe, although the commandeered charge density is of opposite sign, the magnitude is greater than the proton internal density. This destroys the small signal assumption from which it is shown in Chapter 4 that Newton's law of gravitation follows from the new theory. Quite how this affects the interaction of local masses with local masses, and of local masses with distant masses, is not clear. It could be that the analysis of Chapter 4 still applies and therefore local interactions are preserved as predicted by Newton's law of gravitation, as observation requires, but that the effect of distant masses is considerably altered. Further work is required.

17.5 Summary

Table 17.1 Charge densities commandeered within a target particle due to the gravitational potential of massive objects

Position of target particle	Inertial mass of object kg	Gravitational mass of object C	Radius m	Charge density commandeered inside target particle Cm^{-3}
At surface of sun	2×10^{30} (1)	1.9×10^{20}	6.96×10^{8} (1)	2.4×10^{21}
At sun's location in galaxy	2×10^{41} (2)	1.9×10^{31}	2.6×10^{20} (1)	6.5×10^{20}
At centre of universe[6]	(a) 4×10^{52} (4) (b) 2×10^{51} (5)	3.8×10^{42} 1.9×10^{41}	10^{26} (3)	3.6×10^{26} 1.8×10^{25}

(1) Particle Data Group (2014)
(2) Assumes 10^{11} times the inertial mass of the sun enclosed within the sun's galactic orbit
(3) Assumes a nominal radius of 10^{10} light years
(4) Assumes a density of $10^{-26} kg.m^{-3}$ (Bertolami 2006) and a radius of $10^{26} m$
(5) 5% of (a)
(6) Placing all the inertial mass of the universe at the outer radius

17.5 Summary

It is proposed that all space is filled with a background of low density particles. It could be that this accounts for dark matter. Candidates for the background particles are neutrons, hydrogen atoms or electron neutrinos. The background provides a medium for the passage of photons and their transport of energy and gravitational mass. A consequence is that the external gravitational mass and charge densities of particles reduce to low values in the background and this requires that all particles have external coatings of near compensating gravitational mass and charge which nearly cancel the particles' internal gravitational mass and charge. This leads to adjustment of the models for electrons, positrons and electron neutrinos. The charge commandeered in particles due to the gravitational potentials from distant sources is examined and it is concluded that the gravitational

The background

effects from distant objects are expected to be different from local effects described by Newton's law of gravitation.

Chapter 17 references

Bertolami O 2006 arXiv: astro – ph/0608276v2 6 Sep 2006
Particle Data Group 2014 *Review of particle physics* Chinese Physics C Vol 38, No. 9 (2014) 090001

Chapter 18

The origin of the postulates and the fundamental constants

18.1 Introduction

The previous chapters have set out a new theory of physics based on two postulates and five fundamental constants. It starts from within or below particle scale, describes particles, encompasses nuclei and atoms, and extends to massive bodies, to the solar system and maybe to galaxies and beyond. Chapter 13 in the Book examines the possibility of there being no fundamental postulates required to underpin this new theory, and that there are no fundamental constants. It is concluded that the postulates can arise from a process of pure logic, and that the values of the fundamental constants required by the new theory arise from the adopted system of units and measurement. In other words, it is proposed that physics inevitably exists rather than nothing.

The basis of the approach is to appeal to concepts introduced by Descartes and Popper. Descartes proposed that a system of knowledge leading to the laws of physics may be established by logic alone (Descartes translated by Sutcliffe 1968). Popper introduced the concept of falsifiability, that it must be possible for an empirical system, i.e. a theory, to be refuted by experience (Popper 2002 p18). It may be that the work of many philosophers of science is relevant to regarding physics inevitably existing rather than nothing. However here we appeal to the process of pure logic and to the concept of falsifiability. The purpose of this chapter is to re-examine this speculation in the light of the extensions made in the present book, specifically the account in Appendix C and the recasting of the Second Postulate, in support of Chapter 5.

The new theory is based on the two postulates (see Chapter 1, section 1.2):

The origin of the postulates and the fundamental constants

The First Postulate. Each observer observes a three dimensional space filled by two continuous single-valued velocity fields which specify velocity vectors which are functions of time at each point in the three dimensional space.

The Second Postulate. There is a special velocity magnitude such that, when one of the velocity field vectors at a selected point is observed by one observer to have the special magnitude, all observers observe that at the selected point the velocity field has the special magnitude.

However it is pointed out in Chapter 5 of this book, based on the analysis in Appendix C, that observation is in contradiction to the Second Postulate on the following grounds. We observe in the solar system a reduced velocity than c for radio waves when looking tangential to the sun's surface, (see Kenyon 1990 p97) but a local observer will measure c, see Appendix C. A set of postulates which lead to an internal contradiction is not a candidate for the pure logic argument. Hence the original Second Postulate is to be rejected. In Appendix C we propose that the Second Postulate is revised, by adding the word 'local' and by restricting the set of other observers to those in uniform motion with respect to the first observer, to become:

The Second Postulate (revised). There is a special velocity magnitude such that, when one of the velocity field vectors at a selected point is observed by one local observer to have the special magnitude, all local observers in uniform motion with respect to the first observer observe that at the selected point the velocity field has the special magnitude.

In revising the Second Postulate, the mathematics in the Book and in this book remain unchanged. For example, the local analyses involving particles remain unaltered. So do the same conclusions apply regarding the inevitability of physics and there being no fundamental constants? Well - yes. The argument of pure logic leads to the revised postulates, as we demonstrate below. In what follows it repeats the text from the Book, but altered in just a few places to convert the argument to the local form.

We use the expression 'our reality' to refer to our universe if that is all there is, or to the multiverse or, if there is even more, to whatever totality we exist in. In what follows we conjecture, supported by a piece of pure logic, that our reality inevitably must exist. This process does not invoke the anthropic principle that the reality considered has

18.3 The proposed pure logic sequence

to lead to the existence of observers, even though the process does lead to the existence of observers. Specifically we conjecture that a reality based on three spatial dimensions and time and the two postulates must exist

18.2 Something out of nothing

We propose the conjecture that something inevitably exists rather than nothing. Specifically we conjecture that the two postulates above underpinning the new theory, and which we claim determine the nature of our reality, arise inevitably rather than there being nothing. The spur for proposing the conjecture is that the development of the new theory leads to models for fundamental particles. These particles, in one form or another, pervade all space. In free space they are composed of m_{00} and ρ_{00} travelling at speed c so that the finite non-zero quantities $m_{00}/(1 - u^2/c^2)^{1/2}$ and $\rho_{00}/(1 - v^2/c^2)^{1/2}$ result, where $m_{00} = \rho_{00} = 0$. When the particles are in free fall in a gravitational field, the same finite quantities appear out of nothing. Thus the content of the whole of our reality derives from nothing. If we can show that this result, that something exists rather than nothing, can be obtained by pure logic alone, then the new theory derived from the two postulates then follows by pure logic alone. We set out a piece of pure logic as to how this situation can arise – and then we pick holes in it, and the end result is that the initial conjecture remains as conjecture.

18.3 The proposed pure logic sequence

To say there is something is fact empirically, but is an assumption in setting up a theoretical formalism. So start with nothing, and if this inevitably leads to something, then the statement that something exists has been derived theoretically.

So is this nothing confined to zero spatial dimensions? It is a possible description but not a unique description. We can say that at the point x_1, x_2, x_3 there is nothing, or indeed that at each point in the set of all such points, there is nothing.

We can have sequences composed of x_1, x_2, x_3 associated with x_1', x_2', x_3'... associated with x_1'', x_2'', x_3''... etc. at all points of which there is nothing. We can have imaginary dimensions. We can associate each member of the sequence with a new co-ordinate in an imaginary

The origin of the postulates and the fundamental constants

dimension and we can introduce time (strictly all co-ordinates should be made complex, see comments by Penrose (2005 p414)). And why should all these points of nothing be at rest with respect to each other? Thus we can have trajectories in space and time of nothing. These all exist in the totality of nothing – there is no reason to exclude any of these constructs however elaborate. We can therefore have velocity fields where there is a velocity at each point. Within the set of all possible velocity fields, there are those where the velocity at each point is **c**. When c is the same for all local reference frames in uniform motion with respect to each other, then the Lorentz transformation applies, and something, of the form $m_{00}/(1 - u^2/c^2)^{1/2}$ where $m_{00} = 0$ and $u = c$, inevitably appears. This statement is perhaps even stronger – that something can only exist if c is the same for all local reference frames, but this uniqueness needs proof by excluding all other candidates for providing something out of nothing. There can also be velocity fields $\mathbf{c_1}$, $\mathbf{c_2}$, where each has magnitude c at each point and so the reality described by the new theory based on the two postulates inevitably exists. We know that observers can exist in this reality and so inevitably they observe that c is the same for all local observers in uniform motion relative to each other.

18.4 Pure logic versus conjecture

The steps in section 18.3 are in the category of Descartes pure logic leading to conclusions regarding the laws which underpin physics. However there is always a point in such a sequence of pure logic (whether before, at or after the laws have been reached) where a statement is capable of being falsified by the results of experiment and observation. If any statement within the chain of logic could be falsified, then it cannot be deduced to be true by logic alone. Thus a piece of pure logic should contain cast iron steps allowing no alternatives. In not allowing alternatives, it cannot be falsified – and so the piece cannot be a sequence of scientific statements. Conversely, if some or all of the statements are capable of being falsified, then the logic cannot be rigorous. However, as our experience above shows, if the development of the theory from the postulates leads to an inconsistency, then this can be revealed by an experiment which falsifies one of the theory's statements. This follows from the fact that there will be two statements from the theory which are in conflict and

18.6 Summary

an experiment can determine which is false. So we are left with the logical difficulty of not knowing whether there are contradictions within the logical development of the theory. This uncertainty is sufficient to allow for the possibility of falsifiability. On this argument all such attempts at deduction of the postulates by logic alone will be suspect. On the other hand, in so doing they may progress understanding by replacing present postulates by more fundamental ones and maybe economise in their number and content. So with regard to nothing, we are back to where we started – that is, starting with the two postulates, we conclude that our reality is based on nothing (i.e. $m_{00} = \rho_{00} = 0$) and the statement, that the observed physics inevitably exists rather than there being nothing, remains conjecture.

18.5 Fundamental constants

In the Book, Chapter 13, Section 13.6, it is shown that a reality requires no fundamental constants. The set of five constants are replaced by four constants chosen arbitrarily, with no fundamental constants requiring measurement to establish their values. The particular constants chosen to illustrate how the scheme can be set up are the velocity of light, the permittivity of free space, the inertial rest mass of the proton and the ionisation energy of the hydrogen atom.

18.6 Summary

It is conjectured that our reality inevitably exists. It is not clear whether it is unique or whether other realities exist theoretically, even though there may be no means of detecting them. There are no fundamental constants. However there are four constants that can be chosen arbitrarily and their values determine the units for measurement. It would seem that the process of pure logic proposed by Descartes for ascertaining the laws of physics cannot be composed of scientific statements and this process therefore may be impossible. This would also mean that we cannot construct a physics which does not rest on falsifiable a priori postulates. What we can say is that, starting with the two postulates, we conclude that our reality is based on nothing (i.e. $m_{00} = \rho_{00} = 0$).

The origin of the postulates and the fundamental constants

Chapter 18 references

Descartes R *Discourse on method and meditations* Translated by Sutcliffe F E 1968 Penguin Books
Kenyon I R 1990 *General Relativity* Oxford
Penrose R 2005 *The road to reality* Vintage Books (London)
Popper K 2002 *The logic of scientific discovery* 1959 (First English edition), 2002 Routledge Classics (Abingdon)

Chapter 19

The way ahead

19.1 Introduction

This book sets out a new theory of physics. The new theory is based on two postulates and five fundamental constants. A central concept is that the gravitational mass density is equal to the electric potential and the electric charge density is equal to the gravitational potential. Underpinning the development of the theory is a set of Field Equations. Solutions of these equations lead to models for fundamental particles and for photons. Solution of the equations for the case where particles are immersed in potential fields leads to classical and quantum mechanics. Detailed models for various particles are developed. Proposals are made for the origin of the force between nucleons and for the weak force. Many connections are made with the standard model, but there is disagreement in some areas, for example, the standard model distinguishes between quarks of the same flavour by introducing the concept of colour, whereas in the new theory these quarks differ in their x_0 values. Many topics remain to which the new theory can be applied and many details require to be refined, some of which are listed below.

The Book in its Chapter 14 set an agenda for the way ahead. Many of the topics listed for attention have been tackled in the present book. Others remain as unfinished business. So some items below are verbatim from the Book, others are entirely new.

19.2 Accuracy of the predictions of constants and parameters

The prediction in Chapter 8 of the charge on the electron is accurate to better than 20%, and the prediction of the gravitational constant is (fortuitously) even more accurate. However the discussion in Chapter 9 and Appendix D point to a factor of two adjustment in the extent of nucleons along the z axis. The calculations of the various constants and parameters depend on the details of the proton model, and refining the

The way ahead

proton model is a requirement in improving these calculations. A more exact analytical treatment is one possibility. However a more certain route is via numerical solution of the Field Equations. Replacing Figures 9.6 and 9.7 with more exact structures is an important goal.

19.3 Quantum mechanics

In discussing in Appendix F Section F.6 the probabilistic nature of quantum mechanics, it is suggested that there exist oscillatory waveforms in the background which extend over all possible outcomes. These ramp up and become the internal oscillatory charge and gravitational mass densities when an actual particle is present. This speculative proposal needs to be developed into a formal concept within the new theory and then tested against the existing experimental data.

19.4 Quantum field theory and quantum electrodynamics

There are two approaches to the inclusion of quantum field theory and quantum electrodynamics into the new theory. The first, and this is what is done in Chapter 13, is to say that, if the initial starting point for quantum field theory can be derived from the new theory, then the formal content of QFT as conventionally set out then follows, and this leads on to the incorporation of quantum electrodynamics. An alternative and challenging approach is to examine every formal step in QFT in the light of the new theory, to see what changes might have to be made to QFT and to see if some or all of the objections, criticisms, incisive comments or other difficulties that have been advanced, see for example Penrose (2005 pp 675-681), can be resolved. These concern the process of renormalisation, infrared and ultraviolet cut–offs and other issues to do with infinities.

19.5 Nuclear theory

Chapter 11 in the Book is the first foray into models for nuclei. Sufficient has been done to make the overlap mechanism a candidate for the binding of nucleons. However much needs to be done to progress towards a systematic theory of nuclear behaviour.

19.7 Particle decays and interactions

19.6 More particles

Major omissions from the stable of particles predicted by the new theory are the tau and the top quark. The volume lines on the $x-$energy diagram introduced in Chapter 15 Section 15.5 indicate where more particles are to be found. The structures of more mesons and more baryons are expected to be modelled within the new theory, from which predictions of their inertial masses should follow. Again the use of numerical modelling would be a powerful tool.

19.7 Particle decays and interactions

Particle decays and particle interactions need to be described within the new theory. The various symmetries involved need to be accounted for. With hadrons, this is an area where quantum field theory and quantum chromodynamics claim considerable success. The experimental data (Particle Data Group 2014) is vast. However it is not clear to what extent the experimental results are dependent on assuming theoretical models or stand alone irrespective of the current theory. This needs to be untangled before the new theory can make inroads on seeing to what extent the experimental results are predicted from the new theory.

The two approaches of acceptance or challenge can be applied in examining quantum chromodynamics, where a law for the strong potential is assumed and the mechanics of scattering and particle interactions and decays are analysed. The new theory provides the overlap potential and given development of this concept this provides the starting point for examining what can be established from the new theory.

19.8 Electroweak theory

There is no fundamental reason why the new theory should not be extended to cover the topics of electroweak theory or otherwise provide alternatives. However it is difficult to untangle theoretical results in this area which are derived from the formalism of the standard model without further assumptions, and those which are dependent on new assumptions which should be declared as formal postulates. Without this formal structure being made explicit, it is

The way ahead

difficult to decide which parts of the standard model in this area follow from the new theory. This examination of the standard model's formal structure needs to be tackled. In addition within the standard model there is the inclusion of the Higgs mechanism from which it is claimed that the Higgs field is required to imbue particles with inertial mass. The new theory does not need this process since it already has an account of the origin of inertial mass. If the new theory does not require the Higgs mechanism, then how does it account for the CERN observation of a new particle at 125.7 GeV (Particle Data Group 2014)? This all adds up to a large agenda for further work, which as can be seen from Chapter 16 has hardly been started.

19.9 General relativity

The new theory leads to general relativity with various pieces developed in Chapter 5, Appendix C and Appendix 5B in the Book. These need to be brought together into a formal development of general relativity deduced from the new theory. Since modern cosmology is highly dependent on general relativity, it is essential that the form of general relativity is made clear before a serious attempt can be made on cosmology issues.

19.10 The background and cosmology

The unresolved issues and lack of unanimity amongst contributors to the field of speculative cosmology, see for example Penrose (2010), Krauss (2012), signal that there is considerable opportunity for suggestions from the new theory to contribute to the debate. There are many questions. Is there a connection between dark matter and the background? Is there a connection between the energy associated with the travelling oscillatory waveforms in the background and dark energy? What has the new theory to say with regard to the formation of fundamental particles? What is the connection between this and the origin of our universe? More work is required.

19.11 Fundamental constants

It is concluded in Chapter 18, based on Chapter13 in the Book, that there are no fundamental constants. Instead there are four exact numerical values which are selected for the convenience of a particular system of units. The exact numerical values concern the velocity of light, the permittivity of free space, the proton inertial mass and the frequency of a standard clock. The caesium clock is used in the SI system of units to establish the second. In the light of the conclusions from the new theory it is interesting to note that the General Conference on Weights and Measures (CGPM 2014) is deliberating on the proposal that, as well as an exact numerical value for the velocity of light, there be a revision to the International System of Units (the SI) that links the definitions of the kilogram, ampere, Kelvin and mole to exact numerical values of Planck's constant, the charge magnitude of the electron, Boltzmann's constant and Avogadro's constant respectively. The challenge is to make a connection between the two sets of numerical values.

19.12 Tests against experimental data

Above all is the need to test the predictions from the new theory against experimental evidence. The nature of the theory and its results are such that there is ample opportunity to find areas where such tests may be carried out. For example, the theory predicts that particles are not points, but that they are of a size that the details of the predicted structures might be checked experimentally.

Chapter 19 references

General Conference on Weights and Measures *Resolutions adopted by the CGPM at its 25th meeting* (18-20 November 2014)
Krauss L M 2012 *A universe from nothing – why there is something rather than nothing* Simon and Schuster
Particle Data Group 2014 *Review of particle physics* Chinese Physics C Vol 38, No. 9 (2014) 090001
Penrose R 2005 *The road to reality – a complete guide to the laws of the universe* Vintage Books
Penrose R 2010 *Cycles of time – an extraordinary new view of the universe* Vintage Books

Appendix A

The discontinuities in charge density

A.1 Introduction

The purpose of this appendix is to account for how discontinuities in charge and gravitational mass densities can arise, as in Chapter 2 and in Chapter 6. We consider initially how a $1/r$ variation for the charge density ρ_{B1}, bounded by discontinuities, can arise from gravitational mass delta functions.

A.2 The discontinuities

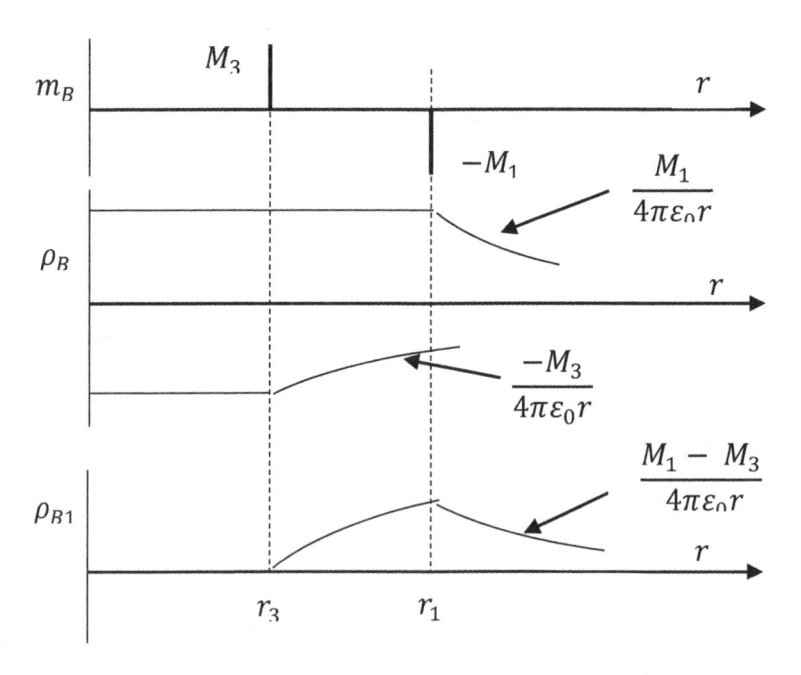

Figure A.1 The spikes of gravitational mass $-M_1$ and M_3 give rise to the charge density variations ρ_B which when summed give the variation ρ_{B1}

A.2 The discontinuities

Consider Figure A.1 with $-M_1$ and M_3 gravitational mass delta functions at r_1 and r_3 respectively. The charge density variations, in their roles as gravitational potentials, arising from $-M_1$ and M_3, are shown, and when combined give the variation shown in the lowest graph. We can arrange that

$$\frac{M_1}{r_1} - \frac{M_3}{r_3} = 0$$

so that the charge density due to M_1 and $-M_3$ cancel for r less than r_3. Define ΔM by

$$\Delta M = M_1 - M_3$$

so that the charge density for r greater than r_1 is given by $\Delta M/4\pi\varepsilon_0 r$. Introduce γ defined by

$$r_3 = r_1(1 - \gamma) \tag{A.1}$$

and so

$$M_1 = \Delta M/\gamma \tag{A.2}$$

and

$$M_3 = \Delta M \left(\frac{1}{\gamma} - 1\right) \tag{A.3}$$

We now keep ΔM and r_1 constant and treat γ as a variable. Equations (A.1) to (A.3) show that r_3, M_1 and M_3 are functions of γ. Let $\gamma \to 0$ and as r_3 approaches r_1 the variation of the charge density becomes a step function, shown at r_1 on Figure A.2, followed by $\Delta M/4\pi\varepsilon_0 r$.

We introduce M_2 at r_2 and $-M_4$ at r_4 with

$$M_1 - M_3 = M_2 - M_4$$

so that the net gravitational mass of the delta functions is zero. We repeat the M_1, $-M_3$ procedure with M_2 and $-M_4$, allowing r_4 to tend to r_2. The thick-outlined region on Figure A.2 is obtained which corresponds to the behaviour of ρ_{B1} in Figure 2.1. This scheme can also be applied to the individual shells introduced in Chapter 6.

The discontinuities in charge density

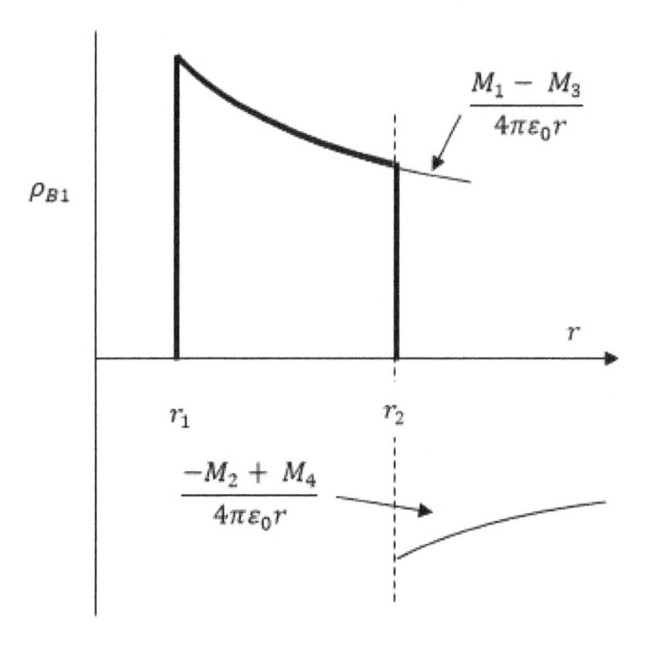

Figure A.2 The charge density variation, shown with a thick outline, resulting from the spikes of gravitational mass $-M_1$ and M_3 at r_1 and M_2 and $-M_4$ at r_2

Of course, in introducing the gravitational mass spikes, we then have the problem as to how we account for them. Consider the 'circuit' shown in Figure A.3 (a) with two capacitances in series, each with capacitance C. $Q = CV$ where V is the voltage across the left hand capacitor. The voltage across the right hand capacitance is $-V$. We let the separation between the capacitances tend to zero producing the situation shown in Figure A.3(b). The integral of the electric potential with x over $2d$ becomes $Vd = M$. Now increase the capacitance by n and the charge by n^2. The distance d reduces to d/n. The peak voltage becomes nV and the integral of the electric potential with x over $2d$ remains Vd, i.e. the gravitational mass remains at M.

A.2 The discontinuities

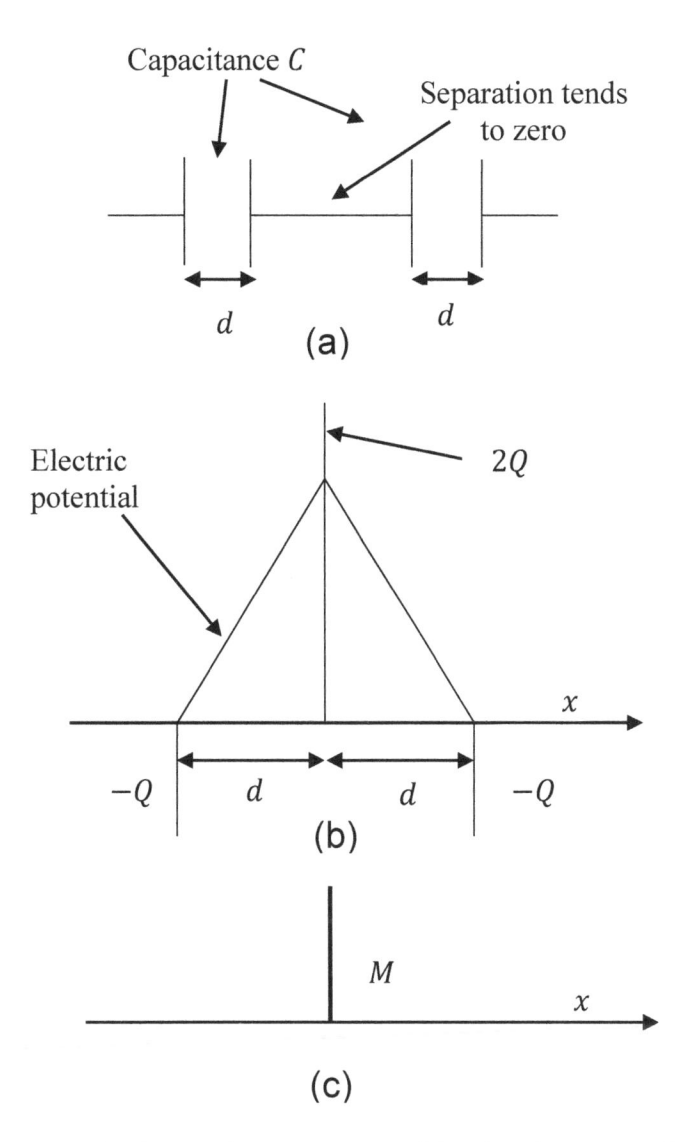

Figure A.3 (a) Charge sheets represented by the charges on the plates of two capacitances connected in series. (b) The charges when the separation distance between the capacitances is reduced to zero. (c) The gravitational mass spike at the centre

Now let n tend to infinity. The separation d tends to zero and the net charge becomes zero but the gravitational mass spike remains at M,

The discontinuities in charge density

Figure A.3 (c). If we allow the charge to increase by a factor n^3, then the product Vd becomes nM, and as n tends to infinity the gravitational mass tends to infinity. Thus we have created the infinite gravitational mass spike used above. The electric charges cancel as d tends to zero.

Now consider the spikes in gravitational mass which give rise to the individual charge spikes Q and $-Q$. By applying a similar process to Q and $-Q$ individually as we have just applied to M, the higher gravitational masses required also cancel, and so on for successive higher components of charge alternating with higher components of gravitational mass. Hence we only need consider the effect of the gravitational mass spikes which give rise to ρ_{B1}, as discussed above. Note that we have used x as the distance variable. In the limits taken, we can replace x by r as used in Figures 2.1 and 2.2

When a Laplace component proportional to $1/r$ is added to the m_B solution, discontinuities are required at r_1 and r_2, and a construction paralleling that for ρ_{B1} applies to m_B.

Appendix B
The omega and photon waveforms

B.1 Introduction

The purpose of this appendix is to revise the mechanism for the propagation of photon waveforms in the background. This leads to a clearer and simplified presentation in Chapter 3 of this book of the material from Chapter 3 of the Book. There is an emphasis on the role of the omega waveform, and a simplification of the propagation in the background of the photon waveform. This allows a clear distinction between the photon waveform in the background and what happens on entry to a target particle.

B.2 The solution of the Field Equations and the omega waveforms

We wish to investigate the nature of travelling waves external to particles and so travelling wave solutions of the oscillatory Field Equations are sought. Since the photon model is to have transverse electric and magnetic fields and similarly transverse gravitational and gravnetic fields, the starting point in Chapter 3 in the Book is to set out the Field Equations in terms of E and G. Because we now wish to emphasise the initial role of the omega waveforms, the new starting point is to use the external fields m_{AE} and ρ_{AE} introduced in Chapter 2 with angular frequency of ω. Thus from Chapters 1 and 2,

$$\frac{\partial^2 m_{AE}}{\partial x^2} + \frac{\partial^2 m_{AE}}{\partial y^2} + \frac{\partial^2 m_{AE}}{\partial z^2} - \frac{1}{c^2}\frac{\partial^2 m_{AE}}{\partial t^2} = -\frac{\rho_{AE}}{\varepsilon_0} \tag{B.1}$$

$$\frac{\partial^2 \rho_{AE}}{\partial x^2} + \frac{\partial^2 \rho_{AE}}{\partial y^2} + \frac{\partial^2 \rho_{AE}}{\partial z^2} - \frac{1}{c^2}\frac{\partial^2 \rho_{AE}}{\partial t^2} = \frac{m_{AE}}{\varepsilon_0} \tag{B.2}$$

We take the direction of travel to be along the z axis. It is shown in the Book, Chapter 3, Appendix 3A, that there exist solutions which are composed of a sum of components in the form of products of a

The omega and photon waveforms

Hermite type function of a carrier described by a sinusoid, and a function of x, y,

$$m_{AE} = \sum_n m_{xyn} a_n h_n(z - ct) \exp(\omega' t - k'_{mn} z)$$

(B.3)

$$\rho_{AE} = \sum_n \rho_{xyn} b_n h_n(z - ct) \exp(\omega' t - k'_{\rho n} z)$$

(B.4)

where the fields are quantised in that the h_n obey the Hermite equation,

$$\frac{d^2 h_n}{dz''^2} - z''^2 h_n = -(2n + 1) h_n$$

This an eigenvalue equation with eigenvalue n with a solution

$$h_n = (-1)^n H_n \exp[-a(z - ct)^2]$$

where H_n is the Hermite polynomial of order n when a function of z'' where

$$z'' = \frac{\omega' z'}{c\sqrt{2n_0 + 1}} \qquad a = \frac{\omega'^2}{2c^2 (2n_0 + 1)}$$

(B.5)

and $z' = z - ct$. $\nu = n - n_0$ and n_0 is a central value of n. For the m components we use

$$\omega'^2/c^2 - k'^2_{mn} = N_{mn} \alpha$$

(B.6)

and for the ρ components

$$\omega'^2/c^2 - k'^2_{\rho n} = N_{\rho n} \alpha$$

where N_{mn} and $N_{\rho n}$ are functions of n and $\alpha = 1/\varepsilon_0$. The solution components in (B.3) and (B.4) apply when $n \gg 1$ and for $\omega^2/c^2 \gg N_{mn}\alpha$ and $N_{\rho n}\alpha$. $h_n(z'')$ is a Hermite function of the parameter z'' given by (B.5) and n determines the number of zero crossings of the function.

In the central region the Hermite function is a sinusoid so that the product with the carrier can be represented by $\exp(i\omega t - ikz)$ where

B.3 Photon waveforms in the background

$\omega = 2\omega'$ where ω is the angular frequency of the source particle. So when the angular frequency is around 10^{24} rad s⁻¹, as in the proton, this is much greater than $(N_{mn}\alpha)^{1/2}c$ which is around 10^{14} rad.s⁻¹ with N_{mn} of the order of unity, and similarly for the ρ terms. So the dispersion effect is negligible. Both envelope and carrier travel at c. These waveforms contain m_{AE} but negligible sinusoidal ρ_{AE} because we have solved equations (B.1) and (B.2) and selected the Type A waveform discussed in Section 3.4 in the Book. Section 3A.5 applies for satisfying the phase condition, equation (3A.11), in the Book. So essentially the omega waveforms travels at c. The mixing zone of the ω_1 and ω_2 waveforms travels at c keeping pace with the omega waveforms.

B.3 Photon waveforms in the background

The photon waveform contains a transverse gravitational momentum component which is proportional to A, the transverse magnetic vector potential due to mixing. This gives rise to the electromagnetic transverse electric and magnetic fields. The photon waveform is part of the $\Delta\omega$ waveform. This is an oscillatory gravitational mass density waveform at $\Delta\omega$ which is accompanied by a steady state gravitational mass density waveform whose integral is ΔM, the transported gravitational mass. This is the mechanism for the transport of gravitational mass by photons.

Somehow the $\Delta\omega$ waveform has to travel at c in the background, both envelope and carrier, in order to keep up with the mixing of the ω_1 and ω_2 waveforms and we must ensure that the phase of the $\Delta\omega$ waveform matches that of the mixing product. Let's introduce another waveform, the $\Delta\omega\, m$ waveform, such that the $\Delta\omega$ waveform contains the gravitational momentum components provided by the $\Delta\omega\, m$ waveform and the photon waveform. We describe the $\Delta\omega\, m$ waveform in similar fashion to the m_{AE} waveforms in the previous section. We start by considering the vector diagrams, Figure B.1, which are discussed in the Book in Sections 3.10 and 12.9. In Figure B.1(a) we show the wave vectors $\Delta k'_{mn}$ and $\Delta\omega'/c$ associated with the carrier and $\Delta\omega'/c$ associated with the Hermite function providing the envelope of the oscillatory m waveform. N_{EBn} is the dispersion factor and is given by

$$\omega'^2/c^2 - k'^2_{mn} = N_{EBn}\alpha$$

The omega and photon waveforms

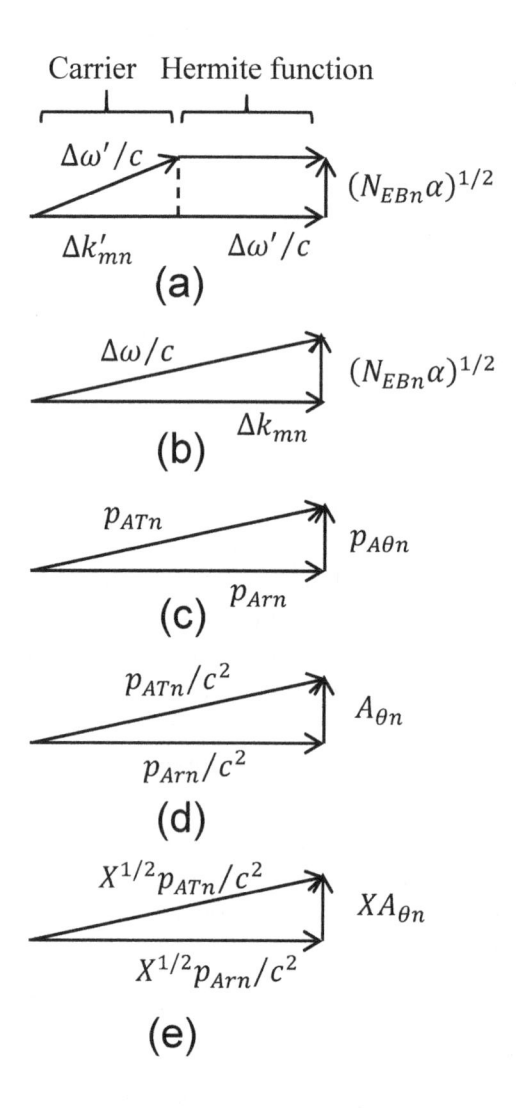

Figure B.1 The m waveform and photon waveform in the background (a) Vector diagram representation of the carrier and Hermite function components for a particular n (b) An approximation to the vector diagram (c) Associated gravitational momentum vectors (d) The equivalent vectors when the transvers vector is the nth magnetic vector potential component (e) The introduction of the factors required in the background where $X = \int_V \psi \psi^* dV$

B.3 Photon waveforms in the background

We simplify the vector diagram by redrawing it as Figure B.1(b). The oscillatory gravitational momentum density of the nth component of the m waveform, p_{Arn}, is proportional to k_{nm} and is in the same direction as k_{nm} where we use spherical co-ordinates consistent with the dipole radiation treated in the Book. If we introduce $p_{A\theta n}$ in a transverse direction, Figure B.1(c), the resultant p_{ATn} can be arranged to be proportional to a wave vector of a waveform, the $\Delta\omega$ waveform, travelling at c in the same direction as the ω waveforms. Thus p_{ATn} has the two orthogonal components

$$p_{Arn} = \frac{m_{Arn} k_{mn} c^2}{\omega}$$

$$p_{A\theta n} = \frac{(N_{EBn}\alpha)^{1/2} c}{\Delta\omega} p_{Arn}$$

where $k_{mn} c^2 / \omega^2$ is the group velocity associated with k_{mn} and m_{Arn} is the amplitude of the oscillatory gravitational mass density. p_{ATn} is the oscillatory gravitational momentum density travelling at c in a longitudinal waveform. The key step is for $p_{A\theta n}$ to be provided by the transverse magnetic potential $A_{\theta n}$, the nth component of the transverse magnetic vector potential, depicted in Figure B.1(d). Summing the $A_{\theta n}$ over n results in the mixing product A. We need to examine what these become in the background and to do this we need to multiply by appropriate factors. Thus we consider the situation at point C, Figure 12.5 in the Book, (Figure 17.1 in this book), not at point B as in Section 12.9 in the Book. p_{Arn} is oscillatory and is multiplied by $\left[\int_V \psi\psi^* dV\right]^{1/2}$ at point C. However because $A_{\theta n}$ is quasi-steady state the factor is $\int_V \psi\psi^* dV$. These are shown in the vector diagram, Figure B.1 (e). The equation in Section 12.9 in the Book becomes

$$\left[\int_V \psi\psi^* dV\right] N_{EB0} R \cong \frac{2 A_m \hbar}{A} \left[\int_V \psi\psi^* dV\right]^2$$

(B.7)

where the factor in square brackets on the left hand side is that applied to the steady state gravitational mass in the background and the factor on the right hand side is that applied to $\Delta\hbar\omega$ in the background.

The omega and photon waveforms

Cancelling by the $\left[\int_V \psi\psi^* dV\right]$ factor obtains the result in Section 12.9 of the Book, but instead of this applying at point B as stated in Section 12.9, it actually applies at point C in the background. Hence $N_{EB0}R \sim 10^{-13}$ in the background and the Book Sections 3A.8 and 3A.9 apply.

Again referring to Figure B.1(e), the transverse component as a fraction of the longitudinal component is considerably reduced in the background compared to the situation in Figure B.1(d) which could apply inside particles (but see the next section). Hence the photon waveform in the background is contained within the longitudinal $\Delta\omega$ waveform travelling with velocity c and which accompanies the omega waveforms. The carrier can be red shifted much closer to zero than with the previous infrared cut-off when Sections 3A.6 and 3A.7 applied.

B.4 Photon waveforms in particles

Within the particle, after removal of the factors discussed in the previous section, from equation (B.7), $N_{EB0}R \sim 3 \times 10^{25}$ and this means that the condition for propagation in Section 3A.8 and 3A.9 in the Book is not satisfied. So ΔM separates from the photon waveform, the latter providing E and H and the transport in or out of energy. We require that G and K are zero, to ensure that energy is not transported by them. We can now apply the sections 3A.8 and 3A.9 in the Book to the propagation of the photon waveform in the basic particle, where it is sufficient to view the event in time only.

B.5 Conclusion

A simplified view is that we need only consider the propagation of the omega waveforms in the background, and thence into particles where there is propagation of the photon waveform with E and H, together with delivery of ΔM. This is the basis of the presentation in this book in Chapter 3. This chapter describes the omega waveforms, the generation of the transverse A, and therefore of E and H, and delivery of gravitational mass, without the detail of the model presented in this appendix for the propagation in the background. This detail means that Table 12.2 in the Book needs correction; a revised version is shown here as Table B.1. Note that we have made the v_0' for the omega

B.6 Comment on Figure 3.8

waveforms distinct from the v_0 for the photon E, H waveforms (also pointed out in Section 3.4), and made other changes and corrections.

Table B.1 The relevant Book Chapter 3 sections together with the entities transported for omega waveforms and photons inside particles and in the background, i.e. at points A and C respectively on Figure 17.1.

	Angular frequency	Distribution parameter	Point A	Point C
Omega waveform m, ρ	$\omega^2 \gg \alpha c^2$	v_0'	3A.5	3A.5
Photons E, H	$\Delta\omega^2 \gg \alpha c^2$	v_0	3A.8	
Photons E, H	$\Delta\omega$ lower than above	v_0	3A.9	
Photons m	$\Delta\omega^2 \gg \alpha c^2$	v_B		3A.8
Photons m	$\Delta\omega$ lower than above	v_B		3A.9
Photon energy transport			$\hbar\Delta\omega$	$\hbar\Delta\omega [\int \psi\psi^* dV]^2$
Photon steady state gravitational mass transport			ΔM	$\Delta M \int \psi\psi^* dV$

B.6 Comment on Figure 3.8

Figure 3.8 does not show the two components of the standing waves at $\omega_1, \omega_2, \omega_3$ and ω_4. There is also the issue that the $\Delta\omega$ waveform accompanies the ω_1, ω_2 overlap away from the source particle but is to

The omega and photon waveforms

accompany the ω_3, ω_4 overlap towards the target particle. We provide an example where these issues are resolved. Consider Figure B.2 with atomic electrons at the foci, F_1, F_2, of a reflecting ellipsoid of revolution about the major axis.

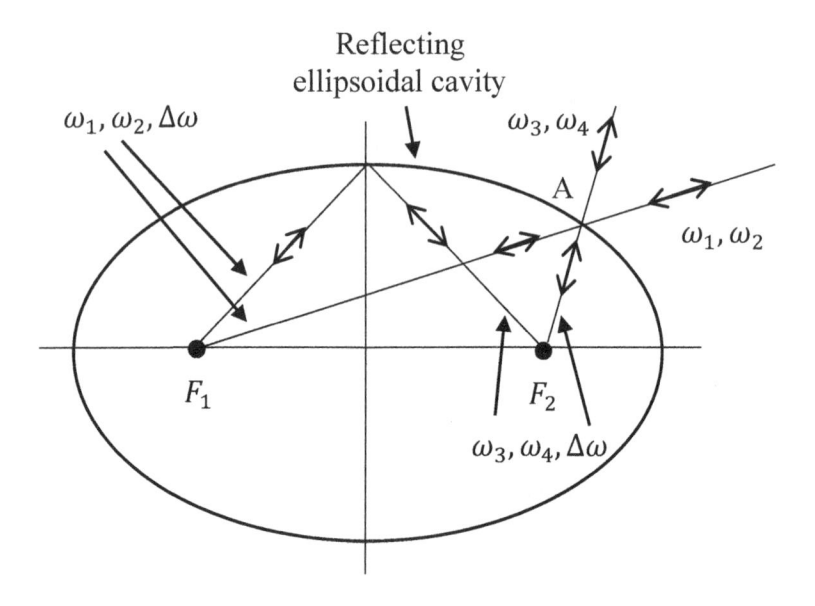

Figure B.2 Two atomic electrons linked by reflections from an ellipsoidal cavity

At A input from free space already exists for the ω_3 waveform as part of the ω_3 standing wave. The $\Delta\omega$ waveform emitted from F_1 is reflected inside the cavity at A and we require this to be accompanied by the overlap of ω_3, ω_4 waveforms towards F_2. This is followed by the emission of the overlap of ω_3, ω_4 waveforms plus the $\Delta\omega$ waveform from F_2 towards A. The reverse process at A results in the overlap of ω_1, ω_2 waveforms plus the $\Delta\omega$ waveform travelling towards F_1. Hence we have the $\Delta\omega$ waveform shuttling back and forth between the two electrons, gaining or losing energy and gravitational mass at each visit. We require to have radial omega standing waves centred on F_1 and which penetrate through the reflecting surface to provide the electric potential due to the F_1 electron, and similarly for the F_2 electron.

B.6 Comment on Figure 3.8

We require two further mechanisms to complete the processes. We initially have the F_2 electron at ω_3 and as previously stated input at A from free space of the ω_3 waveform. Suppose the addition of the $\Delta\omega$ waveform forces the presence of an overlapping ω_4 waveform. On return to A via F_2 the final piece that is missing is the generation of the returning ω_4 waveform from free space. If the propagation of the ω_4 waveform outwards generates a backward wave as suggested in Chapter 3, Section 3.3, then the mechanism for the F_2 standing wave changing from ω_3 to ω_4 is complete. Similar considerations apply to the F_1 standing wave changing from ω_1 to ω_2.

Note that in dealing with an issue arising from the formalism in Chapter 3, we have appealed to the following concepts that have not yet emerged from the formalism: energy, electrons, atoms, a solid cavity, reflecting surfaces, the penetration of solids by omega waveforms, and maybe implicitly many more.

Appendix C

The velocity of muon neutrinos

C.1 Introduction

Adam et al (2011) have described the OPERA neutrino experiment at the Gran Sasso Laboratory in which the velocity of muon neutrinos over the path from CERN was measured to high accuracy. They concluded that the neutrinos arrive at Gran Sasso at an earlier time with respect to the one calculated assuming the velocity of light in a vacuum. The anomaly corresponded to a relative difference of the neutrino velocity with respect to the velocity of light given by $(v - c)/c = (2.48 \pm (0.28 \text{ (stat.)}) \pm 0.30 \text{ (sys.)}) \times 10^{-5}$.

At first I thought that my theory might explain the observed anomaly. However the experimental results were revised and the anomaly evaporated, Adam et al (2012). Thus no longer was there experimental evidence that the particles were travelling faster than light. So when the new experimental results were published, I re-examined my analysis looking for a flaw. I believe the situation is as in this Appendix.

In general relativity it is an inference from the strong equivalence principal that the results of local experiments in free fall are consistent with special relativity (Kenyon 1990 p 12). So general relativity requires that the velocity of light to be c and that particles cannot have a velocity which exceeds this value. With our theory we have to prove that the result follows from the theory and this is the purpose of this Appendix. The analysis is based on the results from Chapter 5 of the Book, which are summarised in Chapter 5 of this book.

The gravitational potential can be defined in two ways either using U based on the gravitational mass or Φ using the inertial mass such that $\mathcal{M}_0 \Phi = M_0 U$ where M_0 is the gravitational mass introduced in Chapter 2 and $\mathcal{M}_0$ is the inertial rest mass introduced in Chapter 4. In this note we use the potential defined using the inertial mass.

C.2 Electromagnetic radiation and particle motion

C.2 Electromagnetic radiation and particle motion

Chapter 3 introduces a model for photon waveforms. The waveform is the product of a sinusoid and a Hermite function. The Hermite function travels at c in zero gravitational potential and this determines the speed of delivery of the photon energy.

It follows from the theory (the Book, Chapter 5, Section 5.4)) that an observer at a large distance from the sun or other massive object, with inertial mass $\mathcal{M}_S$, in the absence of other gravitational sources will observe the group velocity of electromagnetic radiation propagating in the potential of the sun or massive object to be given by

$$c' = c(1 - G\mathcal{M}_S/rc^2)$$
(C.1)

for tangential radiation, and

$$c' = c(1 - 2G\mathcal{M}_S/rc^2)$$
(C.2)

for radial radiation. These results are consistent with extensive observation (for example see Kenyon 1990 p 95 et seq).

There are contributions to the gravitational potential from sources outside the solar system. We require to obtain the effect of these other contributions on the group velocity of electromagnetic radiation. We continue to use a standard clock in zero gravitational potential. Consider a spherical universe with a central hollow sphere. The gravitational mass density may vary with radius, but we assume it is constant with θ and ϕ. The gravitational potential Φ at the centre is a sum of the potentials due to spherical shells and is uniform, i.e. independent of position. Let's now consider the circumstances of the muon neutrino experiment in which they travelled approximately parallel to the Earth's surface. We take the potentials due to the sun and the Earth to be uniform over the neutrino trajectory and included in Φ. Therefore the wave vector is a function of angular frequency, but independent of position, with the result that the group velocity, using (C.1), is given by

$$c' = c(1 + \Phi/c^2)$$
(C.3)

The velocity of muon neutrinos

The same argument can be made for a particle moving with a wave function $\exp(i\omega t - ikz)$ and at the maximum limiting velocity. Consider the dispersion relationship, Chapter 4, Section 4.2,

$$\omega^2 - c^2 k^2 = \omega_0^2$$

When $\omega_0/\omega \to 0$ the velocity approaches the maximum of c. The same analysis in the Book as used for electromagnetic radiation applies to the limiting particle velocity case, and in the gravitational potential Φ the limiting velocity is c' given by

$$c' = c(1 + \Phi/c^2)$$

C.3 Situation with the clock in a gravitational potential Φ

We now consider the case where the standard clock is in a gravitational potential Φ Consider the case where the clock is made using an oscillator which has an angular frequency ω_0 in zero gravitational potential. In potential Φ the angular frequency becomes ω. In Chapter 5, Section 5.4 of the Book, it is shown that

$$\omega = \omega_0(1 + \Phi/c^2)$$

(C.4)

Hence from (C.3) and (C.4) the velocity of light in a uniform gravitational potential is converted from c' to c when using a standard clock in that potential.

Hence it is predicted from the new theory that the velocity of radiation measured using the standard clock in the local potential is c and that the velocity of particles measured using the standard clock in the local potential cannot exceed c, consistent with the revised experimental result.

C.4 Standard clock in free fall

Consider the standard clock initially at rest with the distant observer. We construct the clock so that the clock basic particles have an angular frequency of ω_0 and the steady state gravitational mass and electric charge circulate around the z axis with velocity c. We now allow the clock to encounter a gravitational field and we bring the clock to rest in a uniform potential Φ. The angular frequency, as observed by the

C.4 Standard clock in free fall

distant observer, is given by equation (C.4). The basic particle can interact with the surrounding electromagnetic field and change its state. This requires that the relationship with that field is the same as when the basic particle was in free space. Hence, just as the tangential electromagnetic radiation, when viewed by the distant observer, changes its velocity to that given by equation (C.3), then so must the circulation velocity. This means that the observer at rest with the clock will see an angular frequency of ω_0 and a circulation velocity of c.

Next we allow the distant observer in zero gravitational potential to move. The potential observed by this observer which previously was Φ for the local area becomes Φ'. The observer notes in the local area the radiation velocity is

$$c'' = c(1 + \Phi'/c^2)$$

(C.5)

and the angular frequency becomes

$$\omega' = \omega_0(1 + \Phi'/c^2)$$

(C.6)

and the velocity measured by the local observer is c. Hence for the two local observers moving uniformly with respect to each other, one in Φ and the other in Φ', the velocity of light is c. We can apply this argument to the observation of the circulation velocity which will be observed to be c by other local observers in uniform motion with respect to the initial observer.

We can extend the analysis to the situation where the clock is in free fall down a radius in the vicinity of a massive object. We consider the case concerning the observation of radiation travelling initially horizontally as the observer accelerates from rest in free fall. Text books conclude that radiation entering a free falling box horizontally is observed by the enclosed observer to travel horizontally on account of the radiation bending in a gravitational field. This is also the case with the new theory since we have obtained the bending of radiation result independent of general relativity. At the point of entry the observer's radial velocity is zero and so (C.3) and (C.4) apply and the observed velocity is c. After acceleration, consider the case where the distant observer moves with the local observer's radial velocity. (C.5) and (C.6) now apply and the local observer notes the radiation velocity as c. The same conclusion applies to the circulation velocity when it is

The velocity of muon neutrinos

horizontal, i.e. when the particle z axis is vertical. We can apply (C.5) and (C.6) to a second local observer located with the first local observer but moving uniformly relative to the first observer. This observer also will see radiation and circulation velocities of c.

We next consider radial radiation. The distant observer will see a special relativistic change in the clock frequency, in addition to that given by (C.4), with a factor $(1 - w^2/c^2)^{1/2}$ where w is the free fall velocity. This is the time dilatation effect in special relativity and arises because the clock frequency is displayed alongside the moving local observer and this is then observed by the distant observer. This is in distinction from the observation by the distant observer of the infinite extent travelling wavefunction $\exp(i\omega t - ikz)$ which remains at ω_0 when the particle is in free fall – see Chapter 5, Sections 5.2 and 5.3, and also Chapter 5, Section 5.2, in the Book. Classically

$$\frac{1}{2}\mathcal{M}w^2 = \frac{G\mathcal{M}_s\mathcal{M}}{r}$$

and so

$$(1 - w^2/c^2)^{1/2} = 1 - \frac{G\mathcal{M}_s}{rc^2}$$

and the local clock will appear to have an angular frequency

$$\omega' = \omega_0(1 - 2G\mathcal{M}_S/rc^2)$$

(C.7)

The distant observer notes that the local radial radiation velocity is given by (C.2) and so the local observer in free fall will observe the radiation velocity to be c. The same conclusion applies to the case where the circulation velocity is vertical and therefore is observed to be at c. We can apply (C.2) and (C.7), each modified by a uniform potential shift of $\phi' - \phi$ and equivalent to a change from (C.3), (C.4) to (C.5), (C.6), to a second local observer located with the first local observer but moving uniformly relative to the first observer. This observer also will see radiation and circulation velocities of c.

In the free fall case we conclude that for the two local observers moving uniformly with respect to each other, the velocity of light is c and the circulation velocity is c. Therefore special relativity applies and the local space-time is flat. Since we have satisfied the two postulates of Chapter 1 locally for the free fall case, and since the new

Appendix C references

theory leads to the whole of physics, then the whole of physics applies locally in free fall. This means that we have derived the strong equivalence principle from the new theory.

However there is an anomaly. There is an inconsistency with the Second Postulate in that there is disagreement between the distant observer and the local observers with regard to both the radiation velocity and the circulation velocity. The local observer measures c but the distant observer sees a velocity less than c. This can be resolved if the second postulate has the word 'local' inserted and the set of observers restricted to those in relative uniform motion, and it becomes:

> **The Second Postulate.** There is a special velocity magnitude such that, when one of the velocity field vectors at a selected point is observed by one local observer to have the special magnitude, all local observers in uniform motion with respect to the first observer observe that at the selected point the velocity field has the special magnitude.

Chapter 5 in the Book shows how the general relativity results applying to the solar system follow from the new theory. Appendix 5B in the Book shows that general relativity up to the complexity of the Schwarzschild metric follows from the new theory. Here we have shown that the strong equivalence principle of general relativity follows from the new theory. However for consistency the amendment to the Second Postulate of the new theory is required.

Appendix C references

Kenyon I R 1990 *General Relativity* (Oxford)
OPERA collaboration Adam T et al, 22 Sep 2011 *Measurement of the neutrino velocity with the OPERA detector in the CNGS beam* arXiv: 1109.4897 hep-ex
OPERA collaboration Adam T et al, 12 July 2012 *Measurement of the neutrino velocity with the OPERA detector in the CNGS beam* arXiv: 1109.4897 hep-ex v4

Appendix D

The structure of the proton and related topics

D.1 Introduction

We continue with Set 3 for the proton selected in Chapter 9 of the Book. However it is necessary to delve into the detail of the connection between the spherical shells of the spherical particle model of Chapter 6 and the proton shells of Chapter 9 in order to guide the modelling of the various structures investigated in Chapter 9 of this book and in later chapters and appendices.

D.2 Spherical and cylindrical shells

In the early part of Chapter 6 in the Book, and in Appendix 6A, a spherical particle solution of the Field Equations is investigated. Later in Chapter 6 a model for the proton is introduced, and a comparison is made between the spherical particle solution and that which applies to the proton. The proton solutions are given in Chapter 6, Appendix 6B of the Book and are reproduced here,

$$m_A = \left[\frac{B_{mA}}{r^{1/2}} J_0 \sin^{-1/2}\theta + \frac{B_{mB}}{r^{1/2}} J_1 \sin^{1/2}\theta + C_{mC}\, r^{1/2} J_0 \sin^{1/2}\theta \right.$$
$$\left. + C_{mD} r^{1/2} J_1 \sin^{-1/2}\theta \right] \exp\left(\pm \frac{i\phi}{2}\right)$$

$$\rho_A = \left[\frac{B_{\rho A}}{r^{1/2}} J_0 \sin^{-1/2}\theta + \frac{B_{\rho B}}{r^{1/2}} J_1 \sin^{1/2}\theta\right] \exp\left(\pm \frac{i\phi}{2}\right)$$

(D.1)

These are recast in Section 6.4 of the Book as

D.2 Spherical and cylindrical shells

$$m_A = \left[\frac{B_{mA}}{r_c^{1/2}} J_0 + \frac{B_{mB}}{r_c^{1/2}} J_1 \sin\theta + C_{mC} r_c^{1/2} J_0 \right.$$
$$\left. + C_{mD} r_c^{1/2} J_1 \frac{1}{\sin\theta}\right] \exp\left(\pm \frac{i\phi}{2}\right)$$

$$\rho_A = \left[\frac{B_{\rho A}}{r_c^{1/2}} J_0 + \frac{B_{\rho B}}{r_c^{1/2}} J_1 \sin\theta\right] \exp\left(\pm \frac{i\phi}{2}\right)$$

The Bessel functions are functions of x where $x = \omega r/c$ and these are matched to the spherical case with the adjustment $x' = x - \pi/4$ and with this substitution we can approximate the Bessel functions by expressions proportional to $\sin x'$ and $\cos x'$. A further approximation is to replace r by r_c and Chapter 6, equation (6.9) provides the angle $\Delta\theta$ over which this is valid. Additionally by putting $\sin\theta = 1$ over this angle, the expressions in the square brackets become functions of r_c only. The application of the spherical particle analysis in this case gives rise to cylindrical shells. The differing values for b'_ρ between the spherical particle and the proton cylindrical structure are shown in Table D.1.

Table D.1 Comparison of spherical particle parameters and proton cylindrical structure parameters

Spherical particle		Cylindrical proton		
x	b'_ρ	x	$x' = x - \pi/4$	b'_ρ
7.07	-0.066	7.85	7.07	-0.060
8.64	0.061	9.42	8.64	0.056
10.21	-0.047	11.00	10.21	-0.043

It might be thought that in doubling the size of the nucleon structure by a factor of 2 in Chapter 9 of the Book, because the range in $\Delta\theta$ provided by equation (6.9) has been exceeded, that there are no solutions. Our intent is to show this is not the case. Let's return to equation (D.1). We can extend the analysis to include the effect of $\sin\theta$. Following the procedure using b'_ρ we put

The structure of the proton and related topics

$$\rho_A = B_{\rho A} \sqrt{\frac{2}{\pi x}} \left[\frac{1}{r^{1/2}} \sin x' \sin^{1/2}\theta \right.$$
$$\left. + i(1+b_\rho') \frac{1}{r^{1/2}} \cos x' \sin^{-1/2}\theta \right] \exp\left(\pm \frac{i\phi}{2}\right)$$
(D.2)

leading to, with b_ρ' much less than unity,

$$\rho_B = \frac{2A_\rho B_{\rho A}^2}{r\pi x} \left[\sin^2 x' \sin\theta + (1+2b_\rho') \cos^2 x' \frac{1}{\sin\theta} \right]$$

$$= \frac{A_\rho B_{\rho A}^2}{r} \left[\frac{1}{\pi x} \left\{ \sin\theta + \frac{1}{\sin\theta} \right. \right.$$
$$\left. \left. + \frac{2b_\rho'}{\sin\theta} + \left(-\sin\theta + \frac{1}{\sin\theta} + \frac{2b_\rho'}{\sin\theta} \right) \cos 2x' \right\} \right]$$

Applying the procedure introduced in Appendix 6A of the Book and differentiating with respect to x the expression in the square bracket with respect to x and equating to zero, we obtain

$$-\frac{1}{\pi x^2} \left\{ \sin\theta + \frac{1}{\sin\theta} \right.$$
$$\left. + \frac{2b_\rho'}{\sin\theta} + \left(-\sin\theta + \frac{1}{\sin\theta} + \frac{2b_\rho'}{\sin\theta} \right) \cos 2x' \right\}$$
$$- \frac{2}{\pi x} \left(-\sin\theta + \frac{1}{\sin\theta} + \frac{2b_\rho'}{\sin\theta} \right) \sin 2x' = 0$$

As before we put $\cos 2x' = 0$ and we obtain

$$b_\rho' = \frac{\sin\theta + \frac{1}{\sin\theta} \pm 2\left(-\sin\theta + \frac{1}{\sin\theta}\right) x_0}{-\frac{2}{\sin\theta} \mp \frac{4x_0}{\sin\theta}}$$
(D.3)

$\pm$ is chosen according to whether $\sin 2x'$ is $\pm$ corresponding to b_ρ' being $\mp$ in column 5 of Table D.1. b_ρ' is plotted versus θ on Figure D.1

D.2 Spherical and cylindrical shells

for x_0 equal to 7.85, 9.42 and 11.00. We need to check on the effect of b'_ρ being a function of θ on the $\sin^{-1/2} \theta$ term in the solution (D.2).

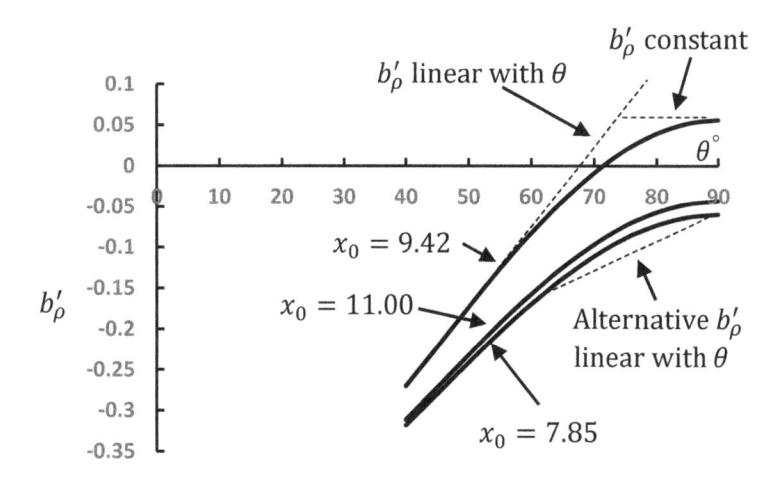

Figure D.1 Plots of b'_ρ versus θ for $x_0 = 7.85$, $x_0 = 9.42$ and $x_0 = 11.00$ using equation (D.3). Each plot may be approximated by a linear section and a constant section, or alternatively by two linear sections

Consider

$$\frac{1}{\sin\theta}\frac{d}{d\theta}\left(\sin\theta\frac{d(b'_\rho\Theta_\rho)}{d\theta}\right) = b'_\rho \frac{1}{\sin\theta}\frac{d}{d\theta}\left(\sin\theta\frac{d\Theta_\rho}{d\theta}\right) + \Theta_\rho\frac{d^2 b'_\rho}{d\theta^2}$$

where $\Theta_\rho = \sin^{-1/2}\theta$. If b'_ρ is either constant or varying linearly with θ, then the second derivative of b'_ρ with θ is zero and the original solution is undisturbed. Figure D.1 shows how the variation with θ can be approximated by a constant section and a linear variation section, or by two linear sections. Choosing these variations for b'_ρ will result in reductions in the shell thickness, see for example Figure 6A.4, Chapter 6 in the Book. Nevertheless we conclude that doubling $\Delta\theta_{3max}$ is justifiable and that a partial spherical form is possible out to

The structure of the proton and related topics

the new $\Delta\theta_{3max}/2$ above and below $\theta = \pi/2$, depicted in Figure D.2.

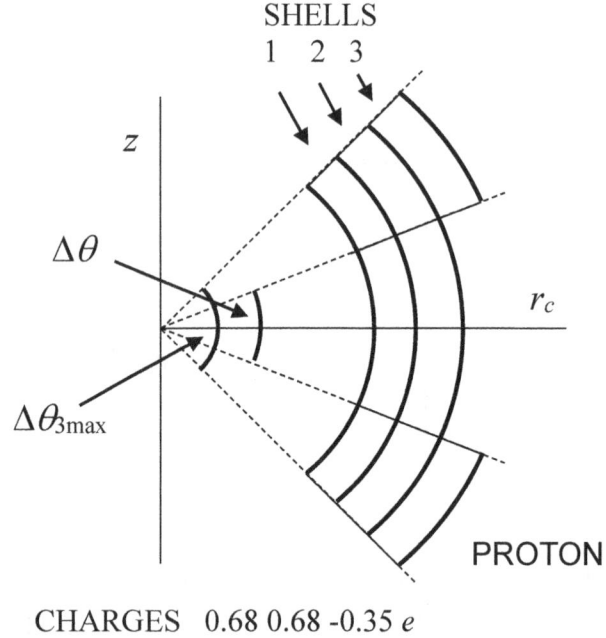

Figure D.2. The proton partial spherical structure with $\Delta\theta_{3max} = 1.52$ rad and $\Delta\theta = 0.72$ rad.

We need to clarify and summarize where we use the various approximations and alternative structures. Sometimes we use the partial spherical form and sometimes we use the cylindrical form. We examine the various cases and why we have chosen the particular form. We deal with the partial spheres first and then the cylindrical structures.

Use of partial spherical configurations. In doubling $\Delta\theta_{3max}$ as discussed in Chapter 9 Section 9.2, we use partial spherical shells. Appendix 9A in the Book is cast in terms of angles (e.g. expressions

D.2 Spherical and cylindrical shells

for η, a'' and b) and can be applied in this doubling process. Data is provided in Table D.2 column (c) (see the next section for more detail) and this is used in the main text beyond Chapter 8 and in the appendices which follow, even when we draw particles as cylindrical structures.

The structure of the proton and related topics

Table D.2 Predicted values of various parameters using (a) the proton model of Chapter 6 (b) the three shell model Chapter 9 and (c) the variant of the Chapter 9 model with $\Delta\theta$ and $\Delta\theta_{3max}$ doubled

	Proton Model Book Chapter 6: Chapter 8 results	Proton Model Book Chapter 9: Set 3 Table 9.3 Book and associated results	Proton Model Book Chapter 9: Set 3 Doubled $\Delta\theta_{3max}$ results
	(a)	(b)	(c)
ω_{0P}/ω_{0e}	1836	345	1836
e C	1.8×10^{-19}	2.21×10^{-19}	1.4×10^{-19}
η	-	0.49	0.98
$\mathcal{G}$ m³ kg⁻¹ s⁻²	7.2×10^{-11}	3.5×10^{-11}	7.0×10^{-11}
$\Delta\theta$ rad	0.7	0.36	0.72
$\Delta\theta_{3max}$ rad	-	0.76	1.52
x_{2P}	13	11.79	11.79
Proton Volume m³	3.0×10^{-44}	1.5×10^{-44}	2.9×10^{-44}
A C s rad⁻¹	1.05×10^{-61}	7.6×10^{-62}	1.1×10^{-61}
A_m C⁻¹ m³	1.7×10^{-2}	7.5×10^{-4}	5.4×10^{-3}
A_ρ C⁻¹ m³	5.0×10^{-25}	6.4×10^{-26}	2.0×10^{-25}
K C² m⁻⁶	-5.87×10^{28}	-1.9×10^{30}	-1.8×10^{29}
a'	-	1.71	1.71
κ	-	1.31	1.31
b	-	0.415	0.415
R	-	1	2.3

Use of cylindrical configuration. The original proton model described in Chapter 6 is limited by equation (6.9). The cylindrical shells extend from $x = 0$ to $x = x_{2P} = 13$. This structure is used in the calculation of

232

D.3 Revised Parameters

parameters in Chapter 8, and data is shown in column (a) of Table D.2.

The proton structure is refined in Chapter 9 in the Book, summarised in Chapter 9 in this book. It results in a structure with cylindrical shells at x_{0i} = 7.85, 9.42 and 11.00. It is depicted in Figure 9.8 in the Book. The data calculated with this structure is shown in column (b) of Table D.2.

However the calculations of the overlap potentials in Chapter 10 use cylindrical shells and these calculations will be more complicated using partial spherical shells. Instead of doubling $\Delta\theta_{3max}$ and other angles, the extent along z of the cylindrical configuration is doubled. The structure is depicted in Figure 9.6 and leads to Figure 10.2.

It is concluded in Chapter 12 Section 12.5 in the Book that the shells in the proton can have motion as independent Dirac particles where energy has been put into the particle which is initially described by a single solution. This concept is modified and extended in Appendix G. We take the proton shell solutions and extract the part around z = 0 (i.e. θ = 90°) and let the extent along z tend to zero, and hence we have a δ function solution. When this is distributed along z by a wave function a cylindrical shell is formed. This construction is used in Appendix G to model some of the excited states of the proton, and modifies and extends the previous analysis.

Also in Appendix G we introduce the concept of a composite particles which contain these independent cylindrical Dirac particles. The cylindrical shape ensures that the Dirac particles can slide past each other without interference and therefore they can have independent oscillatory motions along z. All this leads up to the introduction of quarks. In Chapter 15 a quark is introduced as a δ function structure as described above, with spin ½, and distributed along z to produce a cylinder.

D.3 Revised Parameters

Chapter 8 in the Book obtains the values of many parameters by exploiting the proton model of Chapter 6. Chapter 9 refines the proton structure and Appendix 9A gives the details for obtaining revised parameter values arising from the revised structure. This is a consequence of our use of an analysis of the proton structure as a way of penetrating the dependence of parameters on the values of the fundamental constants. There is a warning in Appendix 9A that the

The structure of the proton and related topics

values of A_ρ and A_m are affected when the choice of proton structure changes. We now attend to obtaining their revised values. We note that A_ρ is determined by finding ρ_B at the edge of the proton and putting $A_\rho = 1/\rho_B$ as in Section 8.5 in the Book. A_m is obtained from A_ρ using the following relationships at the end of Appendix 9A in the Book,

$$R\left(\frac{A_\rho}{A_m}\right)\left(\frac{\omega_{0P}^2}{\alpha c^2}\right) b x_{2P}^2 = 1$$

where

$$b = \frac{\pi(\Delta\theta_{3\max} + \Delta\theta)}{2 x_{2P} \Delta\theta}$$

Using $\rho_A/m_A = \omega_{0P}^2/\alpha c^2$ and equation (9.9) we obtain

$$\rho_B = \frac{q\omega_{0P}^3}{2\pi x_{2P}^3 b \Delta\theta c^3} = \frac{1}{A_\rho}$$

Using this result and the formulae in Appendix 9A we have sufficient means to complete the parameter data in Table D.2. In the main text beyond Chapter 8 and in the appendices that follow we use the parameters in column (c) for the reasons explained in the previous section.

Note that in constructing the set of parameters in Table D.2, column (a) is based on the set of five constants $c, \hbar, \varepsilon_0, \omega_P, \omega_e$ (used initially in order to establish $c, \hbar, \varepsilon_0, A_m, A_\rho$) in Chapter 8. Column (b) is based on $c, \hbar, \varepsilon_0, \omega_P, \mu_P$, which results in Table 9.3 in the Book. Columns (a) and (b) use the condition equation (6.9). Column (c) uses $c, \hbar, \varepsilon_0, \omega_P, \mu_P$ and ω_e and a value of the gravitational constant but in doubling the extent of the particle in θ the constraint of equation (6.9) is dropped as discussed in the previous section..

D.4 The steady state gravitational mass and charge in spherical and cylindrical shells

As explained in section D.2 we apply the same variation of the internal fields to the partial shells and to cylindrical shells. In each shell put (see Book, Appendix 9B)

D.5 Equations for the particle sub-set

$$m_B = m_{BM} x_M^2 / x_{0i} x \tag{D.4}$$

This ensures that when $x = x_{0i}$ (i.e. in the centre of the shell) that the $1/x^2$ variation connecting shells is captured (see Figure 6.1) and then with x_{0i} constant i.e. within a shell, the $1/x$ variation is captured. This means that for spherical shells with thickness $2\Delta x_i$ the gravitational mass in the ith shell, with Δx_i centred on x_{0i},

$$\begin{aligned} M_{i0} &= \int_{2\Delta x} 4\pi x^2 \, m_{BM} (x_M^2 / x_{0i} x) \, (c^3/\omega_0^3) dx \\ &= 8\pi \Delta x_i m_{BM} x_M^2 (c^3/\omega_0^3) \end{aligned} \tag{D.5}$$

and this is proportional to Δx_i. For cylindrical shells with thickness $2\Delta x_i$ the gravitational mass in the ith shell

$$\begin{aligned} M_{i0} &= \int_{2\Delta x} 2\pi x d_i \, m_{BM} (x_M^2 / x_{0i} x) \, (c^3/\omega_0^3) dx \\ &= 4\pi (d_i / x_{0i}) \Delta x_i m_{BM} x_M^2 (c^3/\omega_0^3) \end{aligned} \tag{D.6}$$

With d_i proportional to x_{0i} then M_{i0} is proportional to Δx_i. So for both spherical and cylindrical structures, gravitational mass and similarly the charge in each shell is proportional to the shell thickness.

D.5 Equations for the particle sub-set

Appendix 9A in the Book provides an analysis applicable to the proton. Appendix 9B provides a treatment for neutral particles. Appendix 9B should also apply to the proton in which case we require that equation (9B.3) delivers $\kappa_P = \kappa'_P$. We can satisfy this if the shells are thin and adjacent. Thus we put $x_{0i} = x_{00} + x_i$ where we take x_i to be much less than x_{00} and using $\sum_i C_i = 1$, and after some manipulation, the result follows. Hence it must be implicit in Appendix 9B that shells are thin and adjacent. Although a more general treatment is required, we are going to assume that the thin and adjacent approximation applies and therefore that $\kappa = \kappa'$ applies more

The structure of the proton and related topics

generally. It is concluded in Appendix 9B that for the neutron, and other neutral particles where their charged counterparts are in the subset, the particle sub-set equations apply.

Thus we can use the equations (9A.5) to (9A.8) in Appendix 9A to apply to any particle in the subset, which now includes neutral particles. The full range of particles is summarised in Appendix G. The sub-set equations are obtained using $K = \kappa \rho_{AM} m^*_{AM}$, $M_0 = A\omega$ and $q = A_\rho \rho_{AM} \rho^*_{AM} \mathcal{V}$ and hence

$$|m_{AM}| = \left(\frac{A_\rho K^2 A}{\kappa^2 A_m q}\right)^{1/4} \omega_0^{1/4} \tag{D.7}$$

$$|\rho_{AM}| = \left(\frac{A_m q K^2}{\kappa^2 A_\rho A}\right)^{1/4} \omega_0^{-1/4} \tag{D.8}$$

$$\mathcal{V} = \left(\frac{\kappa^2 A q}{A_\rho A_m K^2}\right)^{1/2} \omega_0^{1/2} \tag{D.9}$$

$$\frac{|m_{AM}|}{|\rho_{AM}|} = \left(\frac{A_\rho A}{A_m q}\right)^{1/2} \omega_0^{1/2} \tag{D.10}$$

We can then obtain the steady state densities using

$$\rho_{BM} = A_\rho \rho_{AM} \rho^*_{AM} \qquad m_{BM} = A_m m_{AM} m^*_{AM}$$

Note that if $\kappa = 1$ the equations above reduce to (6.1) to (6.4). M_0, $\mathcal{V}_0$, m_{BM}, m_{AM}, ρ_{BM}, ρ_{AM}, $|m_{AM}/\rho_{AM}|$ are plotted on Figures D.3 to D.9 versus particle energy for the two values $\kappa = 1$ and $\kappa = 1.31$ (the proton value). The figures in Chapter 8 in the Book are based on Table D.2 column (a). The charts shown here are based on column (c). Using this data and with $\kappa = 1.31$ the volume is given by $\mathcal{V} = 2.782 \times 10^{-56} \omega^{1/2}$. However we know from the table that the proton volume is 2.9×10^{-44} m³. This requires that the volume is given by $\mathcal{V} = 2.43 \times 10^{-56} \omega^{1/2}$ and this means there is an inconsistency in the table, perhaps due to rounding. 2.43×10^{-56} has been used in Figure D.4.

D.5 Equations for the particle sub-set

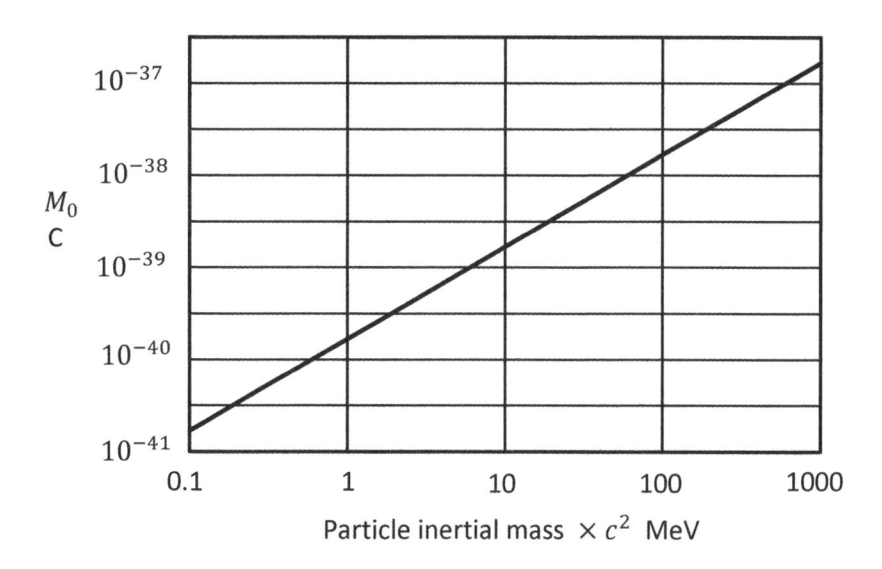

Figure D.3 Particle gravitational mass versus particle inertial mass $\times c^2$ MeV

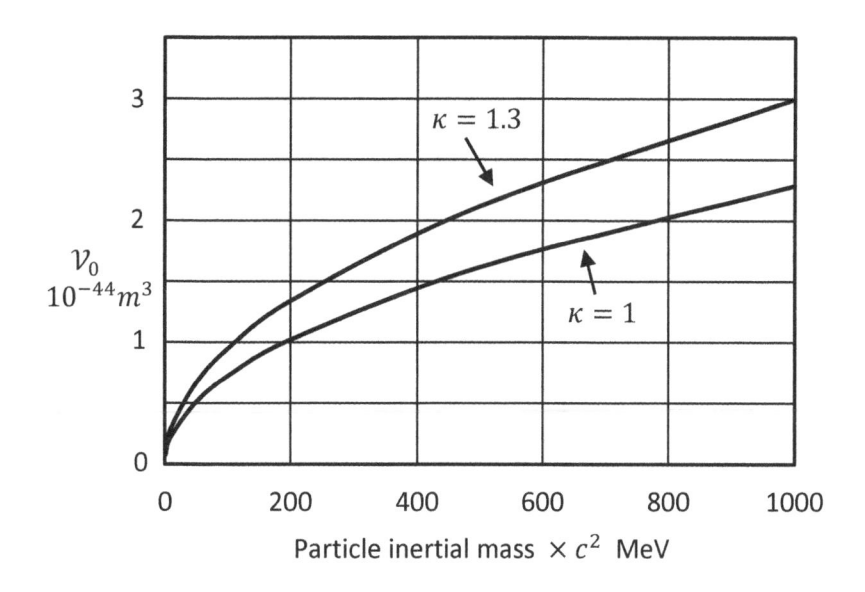

Figure D.4 Particle volume versus particle inertial mass $\times c^2$ MeV

The structure of the proton and related topics

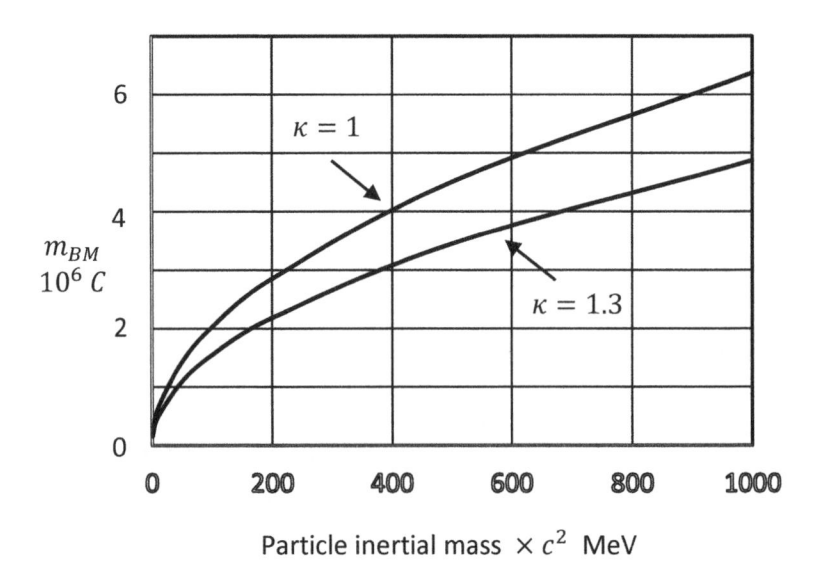

Figure D.5 Steady state gravitational mass density versus particle inertial mass $\times c^2$ MeV

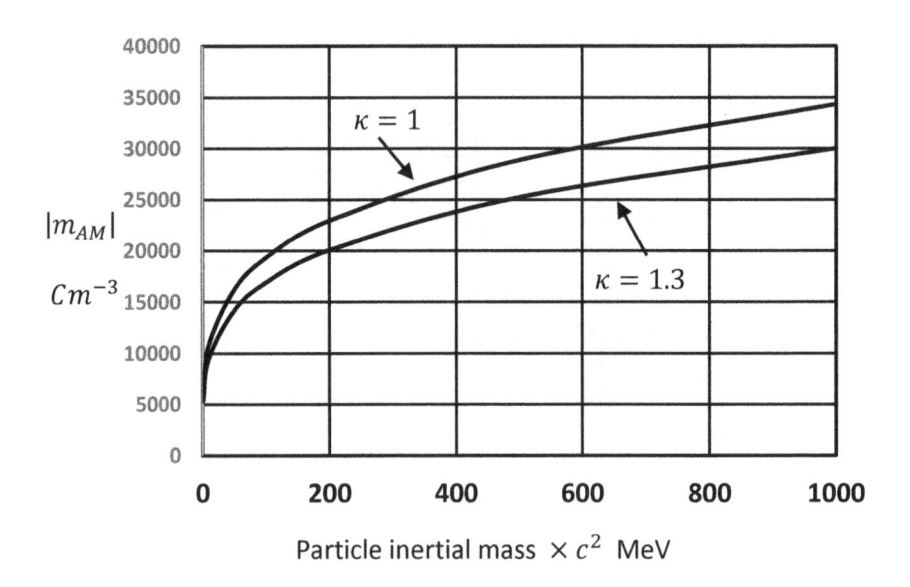

Figure D.6 Oscillatory gravitational mass density versus particle inertial mass $\times c^2$ MeV

D.5 Equations for the particle sub-set

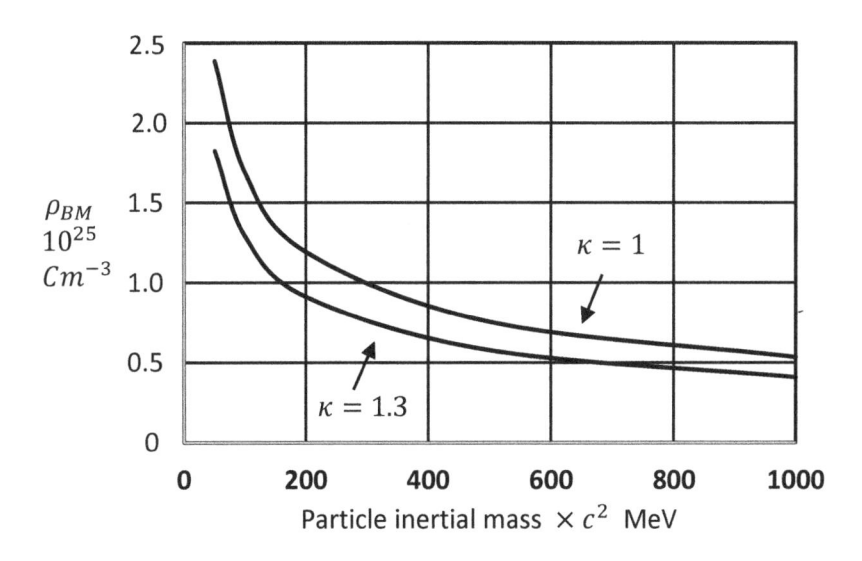

Figure D.7 Steady state charge density versus particle inertial mass $\times c^2$ MeV

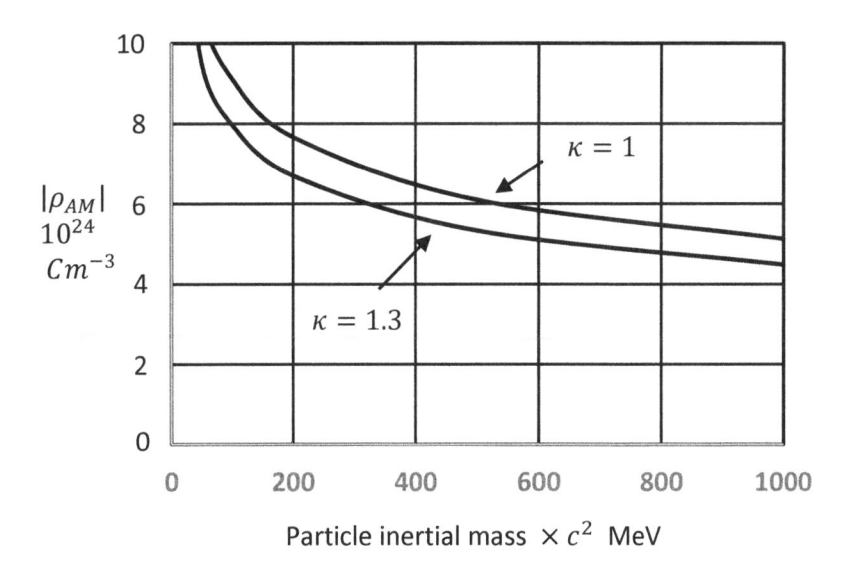

Figure D.8 Oscillatory charge density versus particle inertial mass $\times c^2$ MeV

The structure of the proton and related topics

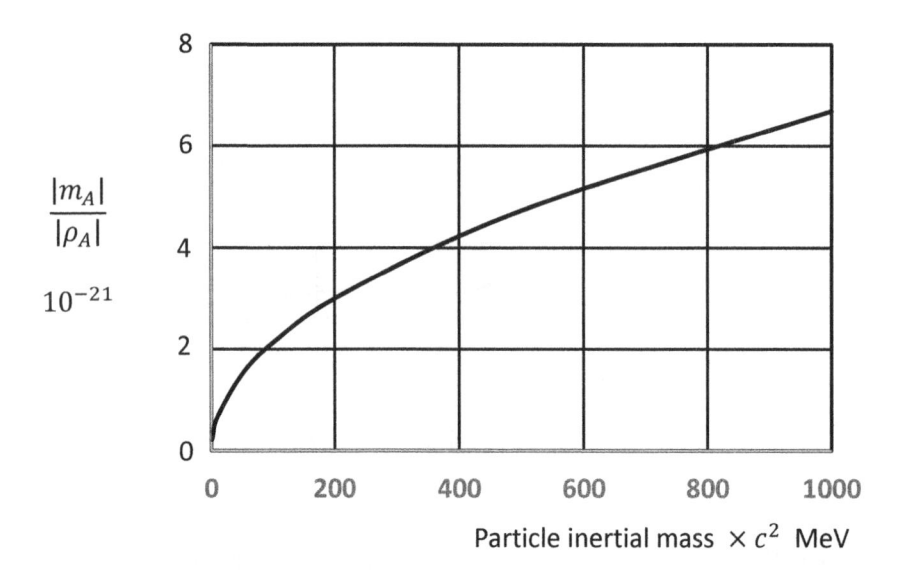

Figure D.9 $|m_A|/|\rho_A|$ versus particle inertial mass $\times c^2$ MeV

Appendix E

The nucleon overlap potential energies

E.1 Introduction

In this Appendix we apply the procedure of Chapter 10 to the overlap of nucleons and so consider a proton interacting with a neutron, a proton interacting with another proton and a neutron interacting with another neutron. This has already been done in the Book, but the procedure is not made completely clear and neither has it been applied accurately. The tables in Chapter 10 of the Book contain errors and as a consequence there are some small errors in some of the potential energy diagrams. Also the plots shown there are against z prior to z being doubled, whereas the interest now is in those plots versus z after it has been doubled, using the proton and neutron models shown in Figures 9.6 and 9.7.

E.2 Nucleon interactions

The procedure is as follows. First we choose z to be in the direction of the 'up' spin of the first nucleon of the pair of interacting nucleons, n-p etc. This means that the circulation parameters identified in Chapter 9 with spin-up apply. Thus when the first nucleon is a proton the circulation parameters are $v_1 = 1$, $v_2 = -1$ and $v_3 = 1$, and when a neutron with spin-up they are $v_1 = 0$, $v_2 = -1$ and $v_3 = 1$. These are reversed when the particles are spin-down. Tables E.1, E.2 and E.3 identify the circulation parameters, the shell charges and the relevant formula for the potential energy from Appendix 10A of the Book.

The nucleon overlap potential energies

Table E.1 The overlap of proton and neutron shells: index to the equations and figures involved. s and m_s describe the spin state. i denotes the shell number. v_P and v_n are the circulation parameters. The interference can be c (constructive) or d (destructive). The overlap parameters are shown in Figure 10.6

s	m_s	Shell	v_P	v_n	c/d	p ±	n ±	PE equation	PE Fig
1	1	1	1	0		+	−		
		2	-1	-1	c	+	+	(10A.12)	E.2
		3	1	-1	c	−	−	(10A.13)	E.2
0	0	1	1	0		+	−		
		2	-1	1	d	+	+	(10A.19)	E.3
		3	1	-1	d	−	−	(10A.20)	E.3

Table E.2 The overlap of proton and proton shells: index to the equations and figures involved. s and m_s describe the spin state. i denotes the shell number. The v_P are the circulation parameters. The interference can be c (constructive) or d (destructive). The overlap parameters are shown in Figure 10.9

s	m_s	Shell	v_P	v_p	c/d	p ±	p ±	PE equation	PE Fig
1	1	1	1	1	c	+	+	(10A.16)	E.5
		2	-1	-1	c	+	+	(10A.16)	E.5
		3	1	1	c	−	−	(10A.17)	E.5
0	0	1	1	-1	d	+	+	(10A.21)	E.6
		2	-1	1	d	+	+	(10A.21)	E.6
		3	1	-1	d	−	−	(10A.22)	E.6

E.2 Nucleon interactions

Table E.3 The overlap of neutron and neutron shells: index to the equations and figures involved. s and m_s describe the spin state. i denotes the shell number. The v_n are the circulation parameters. The interference can be c (constructive) or d (destructive). The shell charges can be positive or negative. The overlap parameters are shown in Figure 10.12

s	m_s	Shell	v_n	v_n	c d	n ±	n ±	PE equation	PE Fig
		1	1	-1	d	–	–	(10A.24)	E.8
		1	-1	-1	c	–	–	(10A.17)	
1	1	1	1	1	c	–	–	(10A.17)	(1)
		1	-1	1	d	–	–	(10A.24)	
		2	-1	-1	c	+	+	(10A.16)	E.8
		3	1	1	c	–	–	(10A.17)	E.8
		1	1	1	c	–	–	(10A.17)	E.9
		1	-1	1	d	–	–	(10A.24)	
0	0	1	1	-1	d	–	–	(10A.24)	(1)
		1	-1	-1	c	–	–	(10A.17)	
		2	-1	1	d	+	+	(10A.23)	E.9
		3	1	-1	d	–	–	(10A.24)	E.9

(1) Overlap of split shells in each neutron is involved. The sequence as the split shells overlap is as follows. (a) The overlap of a lower shell with an upper shell. (b) The overlap of lower and lower shells and the overlap of upper and upper shells. (c) The overlap of an upper shell with a lower shell.

These are followed by the figures showing the potential energy plots. Note that the formula in the Book use q_{Pi} or q_{ni} for the charge magnitudes in each shell whereas these parameters in Chapter 10 in this book are charges which change sign where necessary.

The constants in the various expressions for the potential energy contain the factor

$$\left| \frac{q_{Pi} M_{Pi0}}{V_{Pi0}} \right| = \frac{q_{Pi}^2 A_m}{V_{Pi0} A_\rho} \left(\frac{\alpha c^2}{\omega_{0P}^2} \right)^2$$

243

The nucleon overlap potential energies

From Appendix 9A in the Book

$$\frac{A_m}{A_\rho}\left(\frac{\alpha c^2}{\omega_{0P}^2}\right) = Rbx_{2P}^2$$

and so

$$\left|\frac{q_{Pi}M_{Pi0}}{V_{Pi0}}\right| = \frac{c_{Pi}^2 e^2}{V_{Pi0}}\left(\frac{\alpha c^2}{\omega_{0P}^2}\right)Rbx_{2P}^2 \tag{E.1}$$

where R is the factor introduced in section 9.2, b is defined in Appendix 10A in the Book by

$$b = \frac{\pi(\Delta\theta_{3\max} + \Delta\theta)}{2x_{2P}\Delta\theta}$$

and the c_{Pi} are the proton charge magnitude fractions. In section 9.2 it is suggested that a better fit to the values of major constants can be obtained if the proton's extent (and therefore that of the neutron) along z is doubled, accompanied by a change in R from 1 to 2.3 and e from $e = 2.21 \times 10^{-19}\,C$ to $e = 1.4 \times 10^{-19}\,C$. Equation (E.1) shows that these changes affect the potential energies. The results for the potential energies versus the doubled z_s in the following figures use $R = 2.3$ and $e = 1.4 \times 10^{-19}\,C$.

E.2 Nucleon interactions

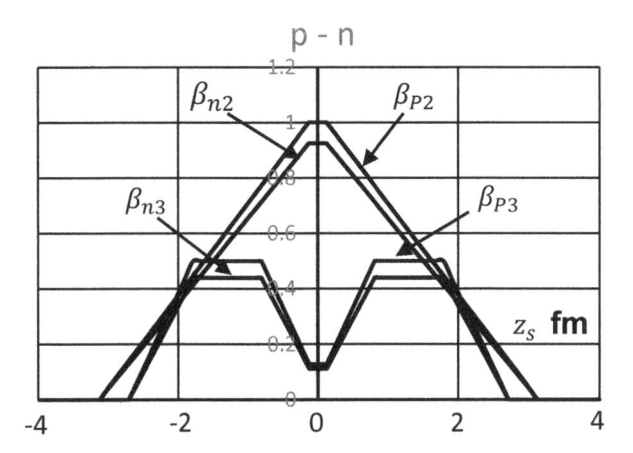

Figure E.1 Overlap parameters β_{P2}, β_{n2}, β_{P3}, β_{n3} versus the separation z_s for p – n interaction, shells 2 and 3

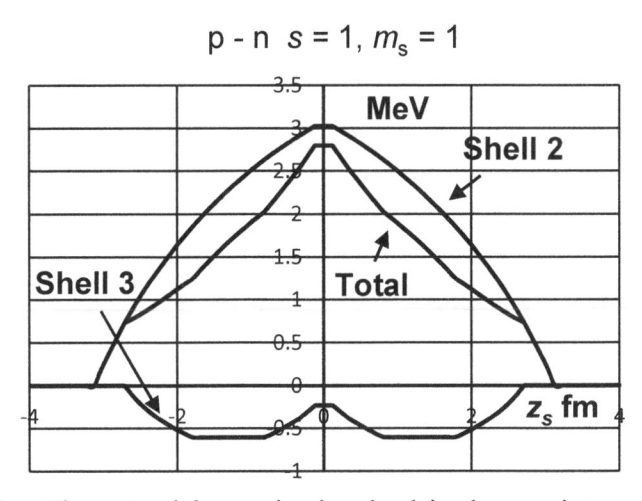

Figure E.2 The potential energies involved in the p n interaction versus separation z_s when $s = 1$, $m_s = 1$

The nucleon overlap potential energies

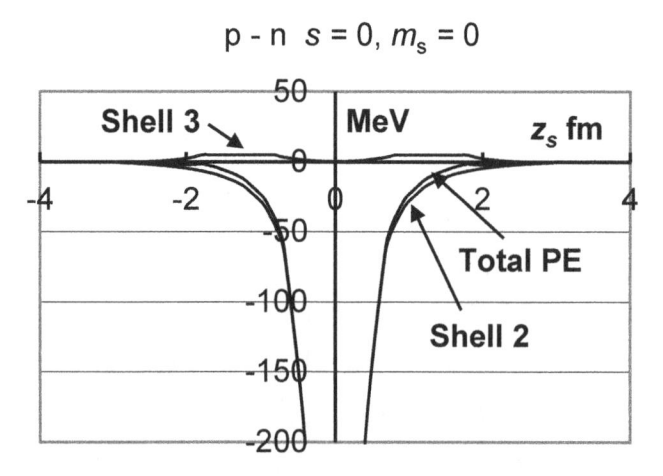

p - n $s = 0, m_s = 0$

Figure E.3 The potential energies involved in the p n interaction versus separation z_s when $s = 0$, $m_s = 0$

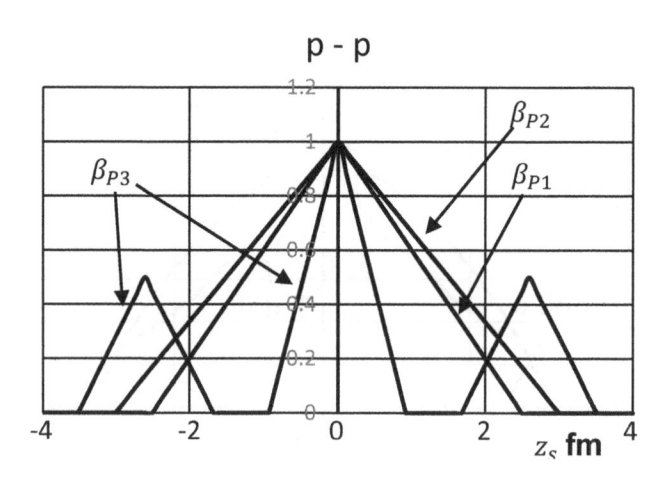

p - p

Figure E.4 Overlap parameters β_{P1}, β_{P2} and β_{P3} versus the separation z_s for p – p interaction, shells 1, 2 and 3

E.2 Nucleon interactions

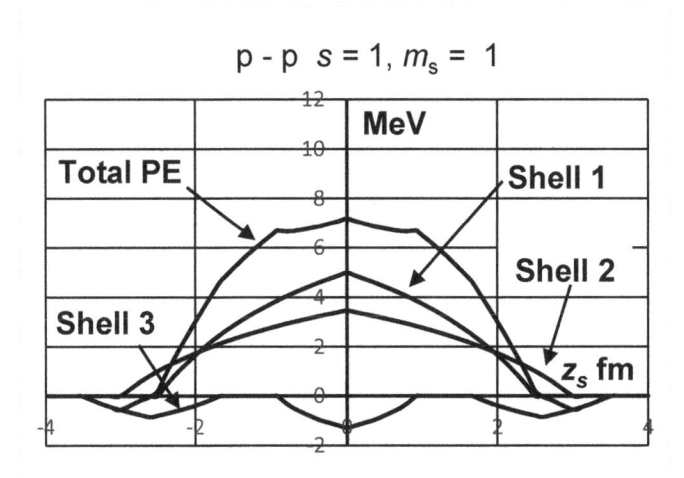

Figure E.5 The potential energies involved in the p - p interaction versus separation z_s when $s = 1, m_s = 1$

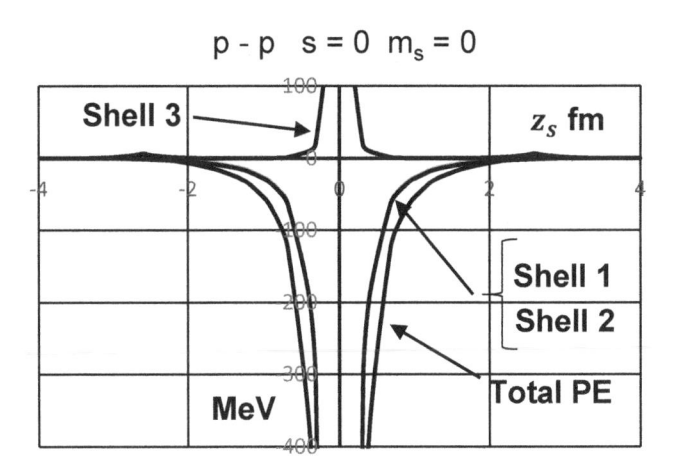

Figure E.6 The potential energies involved in the p - p interaction versus separation z_s when $s = 0, m_s = 0$

The nucleon overlap potential energies

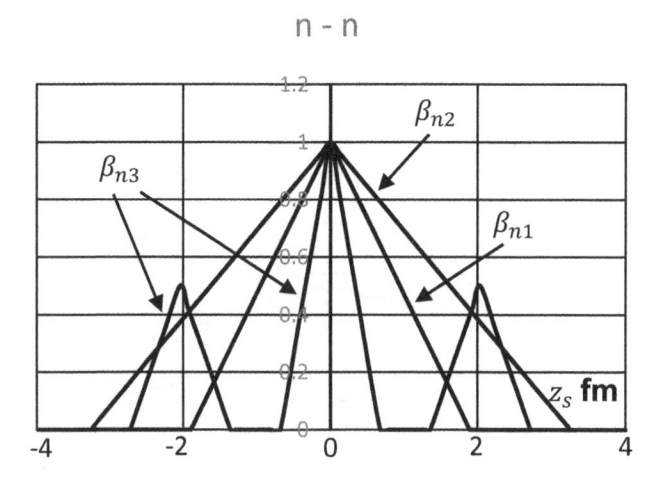

Figure E.7 Overlap parameters β_{n1}, β_{n2} and β_{n3} versus the separation z_s for n – n interaction, shells 1, 2 and

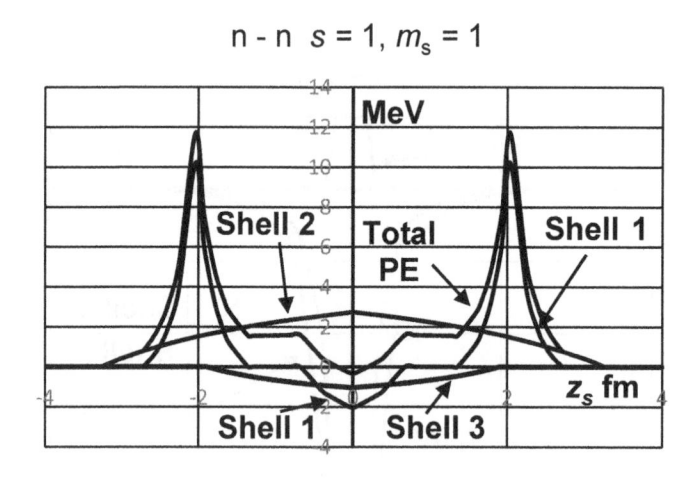

Figure E.8 The potential energies involved in the n - n interaction versus separation z_s when $s = 1$, $m_s = 1$

E.2 Nucleon interactions

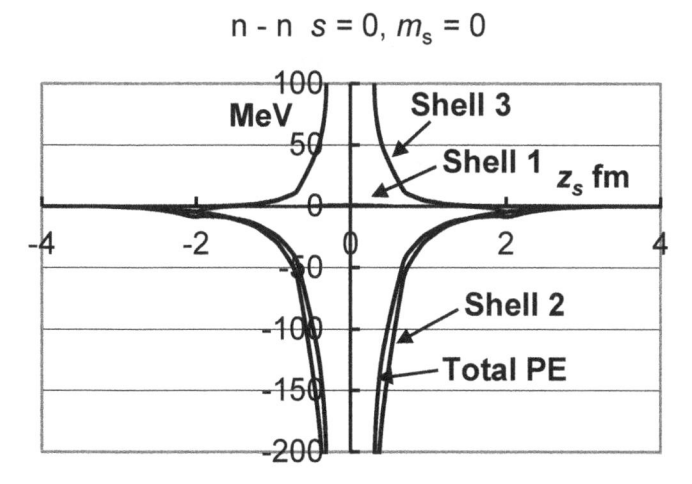

n - n $s = 0$, $m_s = 0$

Figure E.9 The potential energies involved in the n - n interaction versus separation z_s when $s = 0, m_s = 0$

Appendix F

Derivation of the postulates of quantum mechanics

F.1 Introduction

To make the connection formally with quantum mechanics the new theory needs to set out the derivation of the 'postulates of quantum mechanics'. Text books differ considerably on how they present the fundamentals of the subject. There does not appear to be a standard set of postulates and many textbooks do not attempt to list them. Dirac (1981 edition p 15) had the aim that the new scheme becomes a precise physical theory when all the mathematical axioms are specified and that laws are to be stated that connect physical facts with the mathematical formalism. However he does not provide such a list of postulates. Von Neumann (1955) sets out a number of mathematical definitions, and the theorems which follow, on which his mathematical analysis of quantum mechanics is based. However he does not provide an explicit list of physical postulates. Nevertheless we require a starting set of physical postulates. Authors who provide them are not in agreement, though the physical consequences would appear to be the same in the various approaches. The scheme of seven postulates used here is that set out by Dicke and Wittke (1960, p91 et seq), but the wording is also drawn from a number of books and lectures on quantum mechanics over many decades. There is overlap with the five postulates (A, A1, A2, A3 and B) set out by Houston (1959 pp 36-46) and the four postulates stated by Shanker (1994 pp115,116).

The discussion below is based on an interpretation of quantum mechanics stemming from the new theory. The interpretation of quantum mechanics is a major area of debate, with many alternative viewpoints having been proposed and is reviewed in many books, listed in Chapter 12. Our treatment, in which a statistical approach is introduced by choice and resulting in a probability applied to a single

F.1 Introduction

particle, is I think unique, but it is difficult to be sure. It might be said that it is another hidden variable approach, except that the new theory provides a deterministic framework outside of, and underpinning, quantum mechanics. The conclusions from the discussion which follows in Sections F.2 to F.8 is summarised in Chapter 12, Table 12.1.

F.2 Postulate 1

Postulate 1. For a particle moving in an external potential there is a wave function, and all information about the system is to be derived from the wave function.

Discussion. It is established in Chapter 4, via the concept of the particle oscillatory waveform, that there exists a wave function for a single particle, that it is in general complex and that it is required to be a single-valued function of the co-ordinates and time. In the light of the new theory's particle models and the identification of the various fields applying on a scale less than that of the single particle, it does not follow that the wave function determines everything that can be known about the system. The derivation of the various forms of Schrödinger's equation rests on (and therefore assumes) the existence of particles and their properties. This Postulate is claiming unnecessarily more than it need to, and the extra claims are not required in the mathematical manipulation of the wave function.

In the new theory we need to make a distinction between ψ_T and ψ_P. The particle oscillatory waveforms exist as descriptions of the time-varying charge and gravitational mass densities. The wave function ψ_T physically exists and distributes the basic particle oscillatory waveforms. ψ_P also distributes the oscillatory basic waveforms but only as possibilities over the possible range of positions, and introduces probability via $\psi_P \psi_P^*$ as discussed in Chapter 12. In what follows, there is a simplification. In the case of translational motion along z, ψ_T is a function of z and ψ_P is a function of x, y. In the case of radial motion, ψ_T is a function of r, and ψ_P is a function of θ, ϕ. These constructions allow us to write $\psi_T \psi_P$ as the new wave function.

In the case of the bound state, ψ_W exists and distributes the actual charge and gravitational mass of the particle. ψ_P is a function of time and is the coefficient in the form ψ_{Pi} in alternative possible bound states with wave functions $\psi_{Pi}\psi_{Wi}$. ψ_P below refers to either the input

Derivation of the postulates of quantum mechanics

ψ_{P1} to a process or the output ψ_{P2} or both.

F.3 Postulate 2

Postulate 2. For every observable there corresponds an operator. Denote by Q the operator associated with the observable q. The result of a measurement of q is one of the eigenvalues q_n of the eigenvalue equation

$$Q\psi_n = q_n\psi_n \tag{F.1}$$

Case 1. ψ_T, ψ_W. In the discussion below the term 'any operator' refers to any operator in the following list, all from Chapter 4 except that the last one is from Postulate 6 below,

$$i\hbar\,\partial/\partial x, \quad -(\hbar^2/2M_0)\,\partial^2/\partial z^2, \quad (M_0 U_T + qV_T), \quad \mathcal{H}, \quad -i\hbar\,\partial/\partial t$$

and the list is easily extended. By definition the operator satisfies an equation of the form of (F.1). Note that classical mechanics has been derived in the new theory (Chapter 4 of the Book) and we can use results from classical mechanics in examining the fundamentals of quantum mechanics. In general for some physical function F it follows from classical mechanics theory (see Goldstein 1959 p 256 and see below for commentary on the use of Poisson brackets in quantum mechanics),

$$\frac{dF}{dt} = \frac{\partial F}{\partial t} + \{F, \mathcal{H}\}$$

where $\{F, \mathcal{H}\}$ is a Poisson bracket of F with the Hamiltonian. Putting $F = \mathcal{H}$ and, when the Hamiltonian is not an explicit function of time, the energy W is constant. The possible values of W are given by the eigenvalues of $\mathcal{H}\psi_W = W\psi_W$. If the measurement process is applied in a situation where the energy is changing with time but that before or after the measurement process the energy is constant with time, then the constant value of the energy is one of the eigenvalues. A measurement of the energy will give a result which is one of the eigenvalues. An example is the determination of energy level differences using spectroscopy. The measurement of other physical parameters whose operators commute with $\mathcal{H}$ will give a result which is one of their eigenvalues (leading to the identification of 'good'

F.4 Postulate 3

quantum numbers). Another way of expressing this is to say that if a system is an eigenstate of an operator, then we mean that its state is not changing with time. The operator commutes with the Hamiltonian and a measurement on the system will give a result which is an eigenvalue of that operator.

Case 2. ψ_P. $|\psi_P(\theta,\phi)|^2$ describes the probability of finding the particle in the direction (θ,ϕ), say in scattering, i.e. after measurement the particle is in an eigenstate of the direction operator in (θ,ϕ). So Postulate 2 as stated applies.

General comment. Since it is not entirely clear what constitutes 'measurement', it is better to think in terms of processes, with input and output states, as discussed in Chapter 12, and then to analyse specific cases rather than insisting on a generalisation which may be in error. However the general statement that, when a system is in an eigenstate of a particular operator associated with an observable, a measurement will give the eigenvalue of that eigenstate holds in our theory.

F.4 Postulate 3

Postulate 3. Operators associated with physically measurable quantities are Hermitian.

Discussion. It is required that the eigenvalues are real in that the results of measurements with the operators listed above are real numbers. An operator is Hermitian if

$$\int \psi_a^* Q \psi_b \, dV = \int Q^* \psi_a^* \psi_b \, dV$$

where the operator operates on the immediately following wave function. Using

$$Q\psi_n = q_n \psi_n$$

then

$$\int \psi_n^* Q \psi_n \, dV = q_n \int \psi_n^* \psi_n \, dV = q_n^* \int \psi_n^* \psi_n \, dV$$
$$= \int Q^* \psi_n^* \psi_n \, dV$$

showing that the Hermitian condition holds when $a = b$. When $a \neq b$

Derivation of the postulates of quantum mechanics

then applying the orthogonality condition,

$$\int \psi_a^* Q\, \psi_b\, dV = q_b \int \psi_a^* \psi_b\, dV = 0$$

$$\int Q^* \psi_a^* \psi_b\, dV = q_a \int \psi_a^* \psi_b\, dV = 0$$

showing that the Hermitian property holds when $a \neq b$

F.5 Postulate 4

Postulate 4. In general, the set of eigenfunctions of the eigenvalue equation

$$Q\psi_i = q_i \psi_i$$

form an infinite set of linearly independent functions. A linear combination of these functions of the form

$$\psi = \sum_i c_i \psi_i$$

(F.2)

results in an infinite number of possible functions. It is assumed that a physically acceptable wave function can be expanded in eigenfunctions of any observable of the system.

Discussion. The ψ_T and ψ_W form complete sets. The wave function products $\psi_P \psi_T$ form a complete set of ψ_T labelled by ξ and weighted by $\psi_P(\xi)$. The proof of the postulate rests on the wave functions being members of orthonormal sets and since this is the case the postulate follows.

F.6 Postulate 5

Postulate 5. The expectation value of any observable q corresponding to an operator Q resulting from a large number of observations is given by

$$\langle q \rangle = \int \psi^* Q \psi\, dV$$

which can be recast in the form using (F.2),

F.6 Postulate 5

$$\langle q \rangle = \sum_i q_i |c_i|^2$$

(F.3)

This is interpreted by saying that the probability of finding the system in the state i is $|c_i|^2$.

Case 1. If the state is an eigenstate then the expectation value is the eigenvalue. Otherwise we have to consider the process by which the observation is made and Case 2 then applies.

Case 2. ψ_P. As discussed in the main text, we choose to treat processes as statistical physics problems. The range of input and output possibilities can be included in oscillatory waveforms by superposition and this leads to the use of ψ_P in constructing new wave functions covering these possibilities. Steady state values are obtained from the oscillatory waveforms, for example, $\rho_B = A_\rho \rho_A \rho_A^*$. So the resulting wave functions distribute the particle charge and gravitational mass over these possibilities for the purposes of computation. Similarly in forming $\psi\psi^*$ products, probability values are obtained, and sum to unity in the output covering all output possibilities. This means that we have input and output quantum mechanical descriptions which can be connected by quantum mechanical procedures. Thus the output is inevitably in the form $\sum_i c_i \psi_i$ and the expectation value is given by (F.3). This postulate and the interpretation apply to the following three examples.

Scattering. Quantum mechanical analysis of Rutherford scattering using ψ_P is dealt with in Chapter 12.

Spontaneous emission. This is discussed in Chapter 12. The initial state is described by ψ_W and the possible output states by ψ_{W_i}. The presence of the photon waveform has to be treated statistically. ψ_P is a function of time and is the coefficient in the form ψ_{P_i} in alternative possible bound states with wave functions $\psi_{P_i} \psi_{W_i}$.

The double slit experiment with particles. There are two configurations to consider. (a) The interference experiment. We need to ensure that ψ_T has breadth enough to cover the two slits. The ψ_P treatment then accounts for the observed interference statistics. (b) The determination of the particle channel. If we place a detector over the output from one slit, there are two outcomes – either the particle is detected (with a detection probability of 50%) or the particle is

Derivation of the postulates of quantum mechanics

presumed to have traversed the other slit channel. This process is different from that described in (a). The quantum mechanical treatment deals with each case with conclusions consistent with what is observed experimentally. As part of each process we have specified the output states i in (F.3) and the sets of output states differ in the two cases.

The difficulty comes when we say that the input is the same in the two cases, and so how can there be interference if the particle is confined to one slit on each occasion. We say in Chapter 12 and above, that we choose to treat each process as a statistical process. It might be said that we have no choice and that the only way we know is the method of quantum mechanics. We now exercise a choice and enquire whether there is an explanation for the discrepancy between (a) and (b) within the new theory.

Another way of expressing the problem is as follows. While the particle goes through one channel, it would appear that simultaneously there is an oscillatory signal in both channels so that interference can occur at the output from the two slits. We examine this model in more detail.

When the particle goes down one channel, the particle oscillatory waveforms ρ_A and m_A go down that channel, accompanied by the steady state densities. Simultaneously send oscillatory waveforms ρ'_A and m'_A each down both channels with appropriate phases so that interference occurs at the output. For ρ'_A and m'_A not to give rise to steady state waveforms they must be outside of the particle, and therefore in the background (see Chapter 17). These amplitudes are exceedingly small. However when the steady state densities are present in one channel, ρ'_A and m'_A must ramp up from the external values to the internal ρ_A and m_A values. Put

$$\int A_\rho \rho'_A \rho'^{*}_A \, dV = qB$$

$$\int A_m m'_A m'^{*}_A \, dV = M_0 B$$

where the integrals are over all space and

$$B = \int_V \psi \psi^* dV$$

is the background factor from Chapter 17. With

F.7 Postulate 6

$$\int \psi_P \psi_P^* dV = 1$$

we have

$$\rho'_A = B^{1/2} \psi_P * \rho_A$$

with a similar result for m'_A. This gives a physical significance to ψ_P. Maybe there is the possibility of experiments to test these concepts.

This mechanism of sending an oscillatory waveform down both channels, and the steady state down one channel, could provide an explanation for some of the quantum effects observed in the corresponding optical experiments. With both particle and optical phenomena we are appealing to information that we have about the system which lies outside of quantum mechanics, and no amount of discussion within quantum mechanics can access it.

Conclusion. In all three examples $|c_i|^2 = |\psi_{P_i}|^2$ and the interpretation of this postulate follows as shown in Table 12.1.

F.7 Postulate 6

Postulate 6. The development in time of ψ is given by the Schrödinger equation

$$\mathcal{H}\psi = -i\hbar \frac{\partial \psi}{\partial t} \tag{F.4}$$

where the classical expression for the Hamiltonian is used with the quantum mechanical operators substituting for the classical observables.

Discussion. The classical Hamiltonian can have a constant added and Hamilton's equations are still satisfied. However when equation (F.4) is satisfied, it will not continue to be satisfied if a constant is added to the Hamiltonian on the left hand side. There are two cases described below where we can establish the (F.4) result, but there other cases where it does not apply.

Potential well with bound states. Consider the Hamiltonian

Derivation of the postulates of quantum mechanics

$$\frac{-\hbar^2}{2M_0}\nabla^2\psi_W + qV\psi_W + M_0c^2\psi_W = W\psi_W + M_0c^2\psi_W = \mathcal{H}\psi_W$$

(F.5)

from which we write for the rest mass

$$M'_0 = M_0\left(1 + \frac{W}{M_0c^2}\right)$$

(F.6)

Thus the rest inertial mass has become M'_0. This is not made explicit in Section 4.9 in the Book. However the result is generally true when W is introduced, and is used in the frequency balance equation in section 5.2 in Chapter 5 of the Book. Equation (F.6) leads to the bound state angular frequency

$$\omega'_0 = \omega_0\left(1 + \frac{W}{M_0c^2}\right)$$

The time dependent wave function is

$$\psi = \psi_W \exp(i\omega_0't)$$

Substituting for $\mathcal{H}$ in (F.4) results in the correct result for $\partial\psi/\partial t$. However if we drop the M_0c^2 terms in (F.5), resulting in the conventional form of Schrödinger's equation, the value for $\mathcal{H}$ does not satisfy (F.4).

Free particle moving in a potential free region. Introducing time dependence into the wave function we have from Chapter 4,

$$\psi = \psi_0 \exp(i\omega t - ikz)$$

where ψ_0 is a constant and

$$\omega^2 - c^2k^2 = \omega_0^2$$

We can convert this to the relativistic energy – momentum relationship and multiply by the wave function to obtain

$$(\hbar\omega)^2\psi - c^2\mathcal{P}^2\psi = M_0^2c^4\psi$$

We identify $\mathcal{H}$ with $\hbar\omega$ and (F.4) is satisfied.

The general case. We need to take account of the wave function varying due to a time varying amplitude, as well as via a factor of the form $\exp i\omega t$. We have already established classical mechanics, and

F.8 Postulate 7

we require that there is continuity between quantum mechanical behaviour and macroscopic behaviour governed by classical mechanics. Specifically we require that quantum mechanical entities satisfy the following equation from the set of Hamilton's equations,

$$\dot{p}_i = -\frac{\partial \mathcal{H}}{\partial q_i}$$

where p_i and q_i are the canonical variables. Multiply by ψ on both sides and substitute from (F.4) to obtain

$$p_i = i\hbar \frac{\partial \psi}{\partial q_i}$$

(F.7)

which is what we expect as the general form for the momentum operator, confirming the generality of (F.4). As a check we substitute (F.4) and (F.7) into another of Hamilton's equations

$$\dot{q}_i = \frac{\partial \mathcal{H}}{\partial p_i}$$

and it can be seen that this is satisfied. Hence we conclude that (F.4) follows from the new theory. An important consequence is that conventional time-dependent perturbation theory applies with the new theory.

F.8 Postulate 7

Postulate 7. The commutator of the operators Q and R is defined by

$$[Q, R] \equiv QR - RQ$$

and the classical Poisson bracket between observables q and r by

$$\{q, r\} \equiv \sum_i \left(\frac{\partial q}{\partial q_i} \frac{\partial r}{\partial p_i} - \frac{\partial q}{\partial p_i} \frac{\partial r}{\partial q_i} \right)$$

where p_i and q_i are canonical variables. This Postulate requires that

$$|Q, R| = (QR - RQ) \Leftrightarrow -i\hbar\{q, r\}$$

Derivation of the postulates of quantum mechanics

where the observables in the Poisson bracket have been replaced by the corresponding operators. Note that $-i\hbar$ has replaced $i\hbar$ as required previously (see Book Section 4.8). Note also that the co-ordinates and momenta should be with reference to cartesian co-ordinates, see Dicke and Wittke (1960 p103).

<u>Case 1.</u> ψ_T, ψ_W. Substituting from the list of operators in section F.3 above, taking every permutation, the prescription is seen to be satisfied.

Appendix F references

Dirac P A M 1981 *The principles of quantum mechanics* Fourth Edition Oxford University Press
Dicke R H and Wittke J P 1960 *Introduction to quantum mechanics* Addison Wesley
Houston W V 1959 *Principle of quantum mechanics* Dover Publications
Goldstein H 1959 *Classical Mechanics* Addison Wesley
Shankar R 1994 *Principles of quantum mechanics* Second Edition Kluver Academic/Plenum Publishers, New York
Von Neumann 1955 *Mathematical foundation of quantum mechanics* Princeton Landmarks in Mathematics, Princeton University Press, translated from the German by R T Beyer 1949, reprinted 1996

Appendix G

Composite particles and kaons

G.1 Introduction

The purpose of this appendix is to provide supporting detail for the results presented in Chapter 14. A composite particle can be formed by two or more independent spin1/2 Dirac particles in the same particle. The resulting composite particles are cylindrical and they may or may not be stable. They can be the precursor to a single omega solution version of the particle. A key point is that, since the cylindrical structure forms first, this determines some of the properties of the single omega solution, for example its energy. The cylindrical form can have a different decay time and different decay products from the single omega solution.

G.2 Composite particles and the sub-set

Our purpose in this section is to show that a composite particle is in the subset. The sub-set equations are obtained in Appendix D, Section D.5, using $M_0 = A\omega$, $q = A_\rho \rho_{AM} \rho_{AM}^* \mathcal{V}$ and $K = \kappa \rho_{AM} m_{AM}^*$. The first two conditions apply to composite particles but it is required to be shown that the third condition applies. We proceed as follows. We consider charged particles first for which we can write for the ith shell (see the Book, Appendices 9A and 9B),

$$\frac{\omega_i^2}{c^2} a_i q_i = -2\alpha \varepsilon_i A_\rho \rho_{Ai} m_{Ai}^* \mathcal{V}_i + \alpha M_{i0}$$

from which we have

$$\frac{d}{d\omega}(\varepsilon_i \rho_{Ai} m_{Ai}^*) \mathcal{V}_i \frac{d\omega}{d\omega_i} = -\frac{M_{i0}}{2A_\rho \omega_i}$$

If we put

Composite particles and kaons

$$\frac{d\omega_i}{d\omega} = \frac{\omega_i}{\omega}$$

which ensures that $\sum d\omega_i = d\omega$, we obtain by summing over i

$$\frac{d}{d\omega}(\rho_{AM} m^*_{AM}) = -\frac{M_0 \kappa}{2A_\rho \omega V}$$

leading to

$$\hbar = -AA_M K \quad \text{where} \quad K = \kappa \rho_{AM} m^*_{AM}$$

Hence charged composite particles are in the sub-set. For neutral particles we refer to the conclusion in Appendix D, Section D.5, that, given a charged particle is in the sub-set, its neutral counterpart is in the sub-set. Hence we conclude that neutral composite particles are in the sub-set.

There is a further variant we can introduce as follows concerning the annulus shown in Figure 14.3 (a). Starting with the delta function in z and a delta function in r_c, we can convolve the function with r_c with the radial wave function ψ_R introduced in Chapter 4 of the Book, to obtain an annulus. Because ψ_R is common to the whole composite particle, and therefore not specific to each shell, this process creates the same annulus thickness along r_c for all the shells. With a two shell particle, this can only be used when both shells contain the same charge magnitude. So it can be applied to neutral kaons in section G.5 below, but not to charged kaons in Section G.4.

G.3 Kaons

The kaon is dealt with in the Book, Chapter 6 and the particle energy is estimated by assuming that there is continuity of the steady state gravitational mass at the outer radius of a spherical particle, leading to an energy around 330 MeV. This is a lower bound in that there can be modification, by an additional Laplace solution discussed in Chapter 2, which results in a higher angular frequency. In Chapter 12 of the Book the possibility of cylindrical and spherical forms are explored, and these are proposed as the short and long lifetime versions of the kaon. The charged and neutral kaons are explored further below using the concept of the composite particle and with a similar analysis to that used with the proton in Chapter 14, section 14.4, leading to accounts of their

G.4 Charged kaons

constituent charges, particle energies and assessment of their stability.

G.4 Charged kaons

It is predicted that the rest masses of the $K^\pm$ are equal or very nearly so. This is discussed further in Section G.9. We can obtain the individual shell charge fractions by consideration of the cylindrical form, Figure 14.7(a). Each shell has its 'single omega solution' with its own angular frequency. For the reasons explained in Section 6.3 the steady state solutions, although derived from the individual oscillatory solutions, will have a $1/r^2$ variation connecting the shells with local $1/r$ variations in the shells (see Figure 6.1) for the charge and gravitational mass steady state densities corresponding to a steady state single solution at the kaon energy, E. The charge per unit thickness is $2\pi\rho_{B2}r_{02}^2\Delta\theta$ which is independent of r, where r_{02} is the centre radius of the outer shell at which the charge density is ρ_{B2}, see Appendix D, Section D.4. The relative charges in two shells occupied by the independent Dirac particles with the same $\Delta\theta$ are proportional to their x thicknesses at the composite particle energy. At their respective rest mass energies the lower edge is at 0.785, see Chapter 14, Section 14.3, and so the thickness is $2\Delta x = 2(1 - 0.785) = 0.43$. We require this to turn into $2\Delta x_2 = 1.57$ at the lowest composite particle energy in order for the shell to be completely filled. This requires $x_{02} = 3.66$, see Figure G.1 The lower shell is at $x_{01} = 2.09$ to maintain a $\pi/2$ spacing between shells, for which $2\Delta x_1 = 0.90$. The width of the fill of the higher outer shell is 1.75 times that of the lower inner shell and so the charge in the upper shell is 1.75 times that of the lower shell. Since the sum of the charges is $\pm e$, then the outer charge is $+0.64e$ and the inner charge $+0.36e$ for K^+ and the outer $-0.64e$ and the inner $-0.36e$ for K^-. The x_{0i} values are not multiples of $\pi/2$, a situation which can be accounted for as follows. The angular frequencies of these shells are not equal to ω and so it is not necessary that the x_{0i} values to coincide with those of the associated solution.

Composite particles and kaons

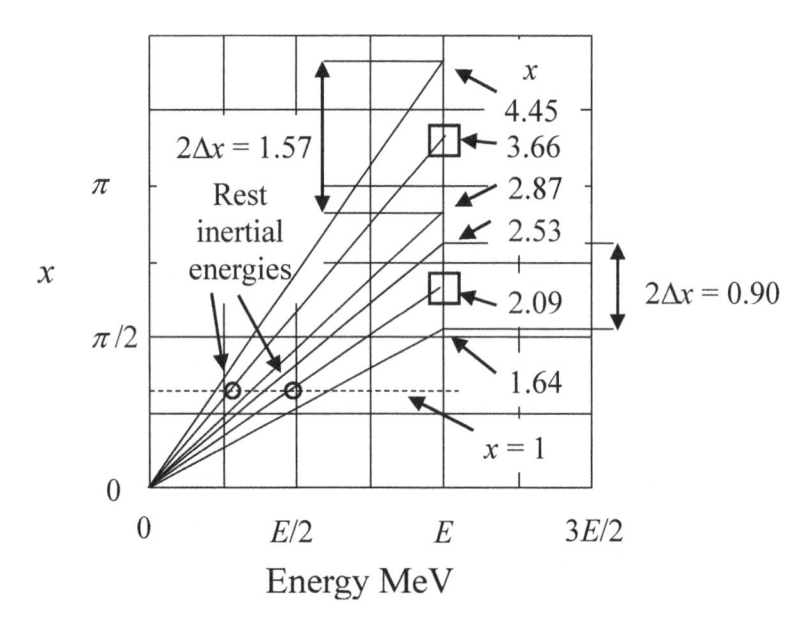

Figure G.1 The composite charged kaon is described by the lower square at $x_{01} = 2.09$ and the upper square at $x_{02} = 3.66$ with shell widths of $2\Delta x = 0.90$ for the lower shell and $2\Delta x = 1.57$ for the upper shell

We now consider motion along z putting $k_i = \pi/2d_i$ as shown in Figure 14.4. Since the composite particle is in the sub-set, the volume has to satisfy, see Section D.5

$$\mathcal{V}_\kappa = \left(\frac{\kappa^2 E}{\kappa_P^2 E_P}\right)^{1/2} \mathcal{V}_P$$

(G.1)

where E is the $K^\pm$ energy and the subscript P refers to the proton. From Appendix 9A, we put $\kappa = 1$ for the $K^\pm$. Now

$$E = (\hbar^2 k_1^2 c^2 + \mathcal{M}_{01}^2 c^4)^{1/2} + (\hbar^2 k_2^2 c^2 + \mathcal{M}_{02}^2 c^4)^{1/2}$$
$$= (1 + x_{01}/x_{02})(\hbar^2 k_1^2 c^2 + \mathcal{M}_{01}^2 c^4)^{1/2}$$

using $d_1 = 2\pi/k_1$, $d_2 = 2\pi/k_2$, $d_2 = (x_{02}/x_{01})d_1$, $\mathcal{M}_{01}c^2 = E/x_{01}$ and $\mathcal{M}_{02}c^2 = E/x_{02}$, leading to

G.4 Charged kaons

$$\frac{d_1\omega_\kappa}{c} = \frac{\pi}{2}\left(1 + \frac{x_{01}}{x_{02}}\right)\left(1 - \frac{(1+x_{01}/x_{02})^2}{x_{01}^2}\right)^{-1/2} \tag{G.2}$$

noting that $d_1\omega_\kappa/c$ is independent of ω_κ. Using (G.1) we have

$$\omega_\kappa^{7/2} = \frac{\kappa_P \omega_{0P}^{1/2}}{V_P} X$$

where the volume is given by

$$V_\kappa = 4\pi\left(x_{01}\Delta x_1 + \frac{x_{02}^2 \Delta x_2}{x_{01}}\right)\left(\frac{d_1\omega_\kappa}{c}\right)\left(\frac{c^3}{\omega_\kappa^3}\right) \equiv \frac{X}{\omega_\kappa^3} \tag{G.3}$$

where $2\Delta x_1 = 0.90$ and $2\Delta x_2 = 1.57$ and this leads to a predicted energy of $E = 507$ MeV. However the cylindrical form of $K^\pm$, although matched to the spherical particle fields, will have considerable leakage through upper and lower surfaces as discussed in section 9.2 and so will convert to the 'single solution' spherical form. Note that $E_1 = 323$ MeV, $E_2 = 184$ MeV and these energies are indicated on Figure 14.4.

In chapter 8 of the Book, the charge density at the outer radius of the proton is $1/A\rho$. We introduce some additional notation for the kaon. r_{11} and r_{12} refer to the inner and outer radii of the inner shell, and r_{21} and r_{22} refer to the inner and outer radii of the outer shell. There is corresponding notation for the x values. Inspecting the variation of ρ_{BM} with angular frequency, see Figure D.7, $\rho_{B22} > 1/A\rho$ for particles of lesser energy than the proton. However $\rho_{B22} < 1/A\rho$ for particles of greater energy than the proton, suggesting instability. These conclusions rest on $\kappa_P = 1$ and $\kappa = 1$ but the conclusions remain for significant departures from the proton energy. Now

$$\rho_{B22} = \left(\frac{e}{V_\kappa}\right)\frac{x_M^2}{x_{02}x_{22}}$$

where the construction involving $x_M^2/(x_{02}x_{22})$ is explained in Appendix 9B in the Book.

For $K^\pm$, $\rho_{B22} = 5.3\times10^{24}$ C cm^{-3} and this is greater than $1/A\rho = 4.93\times10^{24}$ C cm^{-3} as required. Data is in Table G.1.

Composite particles and kaons

Table G.1 Comparison of the charged and neutral kaons predictions – cylindrical structures

Particle form		$K^{\pm}$	$K^0, \overline{K^0}$
Cylindrical			
	Predicted energy MeV	507	513
	E_1 MeV	323	321
	E_2 MeV	184	192
	x_{01}	2.09	2.36
	x_{02}	3.66	3.93
	r_{01} fm	0.81	0.91
	r_{02} fm	1.42	1.51
	d_1 fm	1.46 fm	1.32
	d_2 fm	2.55 fm	2.19
	Volume V_1 m³	2.6 × 10⁻⁴⁵	4.5 × 10⁻⁴⁵
	Volume V_2 m³	1.39 × 10⁻⁴⁴	1.26 × 10⁻⁴⁴
	Particle volume m³	1.65 × 10⁻⁴⁴	1.71 × 10⁻⁴⁴

For the proton, continuity between the internal m_B and the external electric potential was assumed, see Chapter 6. For $K^{\pm}$ the electric potential at r_2 is given by

$$V = \frac{e}{2\pi\varepsilon_0 r_2 \Delta\theta_\kappa}$$

(G.4)

where $\Delta\theta_\kappa = d_1/x_{01} = d_2/x_{02}$ and hence

$$x_M^2 = (x_{01}^2 + x_{02}^2)/2$$

so that

G.4 Charged kaons

$$m_{B22} = \left(\frac{e}{V_\kappa}\right)\frac{x_M^2}{x_{02}x_{22}}$$

m_{B22} is about $4V$ where (G.4) is used to calculate the external electric potential. The discrepancy can be reduced if R (see Appendix 9A in the Book) is introduced, though a discontinuity in the steady state gravitational mass density may remain at the boundary.

For the spherical single omega solution form of the charged kaon, the maximum volume of the two shells is given by,

$$V_1 = \frac{4\pi}{3}(x_{12}^3 - x_{11}^3)\left(\frac{c}{\omega_\kappa}\right)^3$$

$$V_2 = \frac{4\pi}{3}(x_{22}^3 - x_{21}^3)\left(\frac{c}{\omega_\kappa}\right)^3$$

and the maximum total volume is the sum of these. Table G.2 shows that there is sufficient volume available in the spherical shells to accommodate the charge and gravitational mass of the cylindrical form of the particle.

The spherical solution is constrained by the volume condition at ω, equation (G.1). The available volume exceeds the required volume and this results in fractional shell filling, see Table G.2. The spherical shell thicknesses are calculated as follows. The thicknesses are in the ratio of the charge fractions, see Appendix D, Section S4,

$$\frac{2\Delta x_{s1}}{2\Delta x_{s2}} = \frac{c_1}{c_2}$$

where the s subscript refers to the spherical particle. The spherical particle volume is equal to $\mathcal{V}_\kappa$ and so

$$\left(4\pi x_{s01}^2 \frac{c_1}{c_2} + 4\pi x_{s02}^2\right)2\Delta x_{s2}\left(\frac{c^3}{\omega_\kappa^3}\right) = \mathcal{V}_\kappa$$

(G.5)

and this determines Δx_{s2} and then Δx_{s1}. The values are in Table G.1. Whereas the K^0 has two forms, discussed in the next section, the $K^\pm$ has only the spherical long lifetime form.

Composite particles and kaons

Table G.2 Comparison of the charged and neutral kaons predictions – spherical structures

Particle form		$K^{\pm}$	$K^0, \overline{K^0}$
Spherical			
	Predicted energy MeV	507	513
	x_{s01}	2.36	2.36
	x_{s02}	3.93	3.93
	r_{s01} fm	0.92	0.91
	r_{s02} fm	1.53	1.51
	Available $\mathcal{V}_1$ m^3	6.7×10^{-45}	6.5×10^{-45}
	Available $\mathcal{V}_2$ m^3	1.82×10^{-44}	1.75×10^{-44}
	Available particle volume m^3	2.49×10^{-44}	2.40×10^{-44}
	$2\Delta x_{s1} c/\omega_K$ fm	0.27	0.44
	$2\Delta x_{s2} c/\omega_K$ fm	0.47	0.44
	r_{s11}	0.78	0.69
	r_{s12}	1.05	1.13
	r_{s21}	1.29	1.29
	r_{s22}	1.76	1.73

G.5 Neutral kaons

G.5 Neutral kaons

There are the two forms, the cylindrical and the spherical, introduced in Chapter 12 of the Book. The cylindrical form is formed first and determines the energy and so we start with it first. We make the x_{0i} levels of the Dirac particles at the kaon energy in the cylindrical form correspond to the x_{0i} levels of the spherical shells, i.e. $x_{01} = 3\pi/4 = 2.36$ and $x_{01} = 5\pi/4 = 3.93$ and hence facilitate the transformation into the spherical form. There will be the same magnitude of charge in each shell and so they will need to have the same thickness, but we do not yet know this thickness or the charge fractions $c_1 = c_2$. With the cylindrical form, with superposition of solutions with a range of displacements of the origin of r_c for both Dirac particles using a convolution with ψ_R as discussed in Section G.2 above, then the two shells can be filled out to any required thickness up to $\Delta x = \pi/2$. We are taking the shells to be filled, i.e. $2\Delta x_2 = 2\Delta x_1 = 1.57$, see Figure G.2. There is motion of the shells along z as with $K^{\pm}$ above.

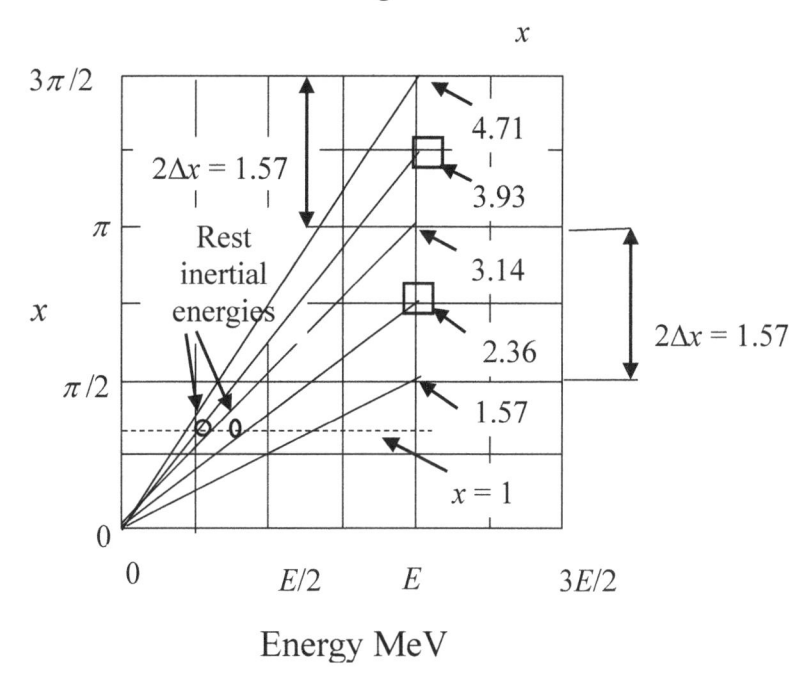

Figure G.2 The composite neutral kaon is described by the lower square at $x_{01} = 2.36$ and the upper square at $x_{02} = 3.93$ with shell widths of $2\Delta x = 1.57$ for both shells

Composite particles and kaons

Again we apply the principles for composite particles. The shells will have the same steady state solutions as an energy E single solution but the oscillatory waveforms are at the lower angular frequencies of the two Dirac particles. The spherical form has a single solution as stated in Chapter 6 and the same energy as the cylindrical form. This will determine its volume, but its shells will not be completely filled, as we show below.

Equation (G.1) applies, but we do not know κ. We can apply (G.2) above to determine that $d_1 \omega_K/c = 3.42$ where ω_K is now the neutral kaon angular frequency corresponding to its energy E. Equation (G.3) applies with $\Delta x_2 = \Delta x_1 = \pi/4$. The $K^\pm$ analysis shows that at the outer radius of the cylindrical structure the charge density is nearly equal to $1/A_\rho$ and so we shall put

$$\frac{1}{A_\rho} = \frac{qx_M^2}{V_K x_{22} x_{02}}$$

where $q = \kappa^2 e$ and V_K is the volume. Thus we have

$$\kappa = \left(\frac{V_K x_{22} x_{02}}{e A_\rho x_M^2}\right)^{1/2}$$

So

$$\omega_K^4 = \frac{X e x_M^2 \kappa_P^2 \omega_{0P}}{x_{22} x_{02} V_P^2}$$

The predicted energy for the neutral kaon is 513 MeV. $E_1 = 321$ MeV and $E_2 = 192$ MeV. $x_M^2 = 10.49$, $\kappa_K = 1.03$ and c_1 and $c_2 = 0.53$. The volume is 1.7×10^{-44} m³. Data is in Table G.1.

For the spherical form the filled volume over the two shells is given by

$$\frac{4\pi c^3}{3\omega_K^3}(x_{22}^3 - x_{21}^3 + x_{12}^3 - x_{11}^3)$$

and at 2.4×10^{-44} there is sufficient to accommodate the volume required Since $c_1 = c_2$, the actual shell thicknesses required are equal and are determined by (G.5). Detail is provided in Table G.2.

270

G.6 The inertial masses of particles and their antiparticles

G.6 The inertial masses of particles and their antiparticles

The purpose of this section is to explore the consequences of some aspects of anti-particles with positive gravitational mass. The first point concerns the electric charge. For the positively charged particle the integral of the steady state charge density $A_\rho \rho_A \rho_A^*$ over the particle volume is $\pm e$. For the anti- particle if we have the same oscillatory waveform, the same angular frequency and the same volume and so with $-A_\rho$ the electric charge is $-e$. The situation for the gravitational mass is more complex. We investigate it using the analysis of the spherical particle in Chapter 6 of the Book. Consider the right hand side of equation (6A.4) in Appendix 6A in the Book; circumstances are identified under which this term can be neglected. A more general approach is as follows. Using equation (6.10) in the Book, $A_\rho = A_m \alpha c^2 / \omega_{0P}^2 x_{2P}^2$ and from Section 6.3 in the Book $f^2 = \omega_0^5/\omega_{0P}^5$ then

$$\alpha A_\rho \rho_A \rho_A^* = \frac{A_m m_A m_A^* \omega_{0P}^3}{x_{2P}^2 \omega_0 c^2}$$

which is small compared to $2A_m m_A m_A^* \omega_0^2/c^2$ (the negative of one of the terms on the left hand side of (6A.4)) for $f \geq 1$. This means that the Laplace solution is a good approximation for the steady state gravitational mass density and so there is little to distinguish the antiparticle steady state gravitational mass density solution from that of the particle solution. So given the oscillatory solution for the positive e particle, then this also applies to the anti- particle.

We can investigate further the difference between the two steady state solutions as follows. Because of the shell structure with discontinuities between them, we do not need to consider an extra continuous Poisson contribution. Instead we consider the change in gradient with r from the centre of a filled shell to the edge due to the Poisson component which is

$$\frac{A_m m_A m_A^* \omega_{0P}^3}{x_{2P}^2 \omega_0 c^2} \left(\frac{\pi c}{4\omega_0} \right)$$

The Laplace solution has a gradient $-A_m m_A m_A^* \omega_0 / x_0 c$ and the Poisson change is small compared to the magnitude of this expression.

We conclude that the particle and antiparticle share the same

Composite particles and kaons

oscillatory charge density solutions, but with change in sign of the steady state charge density solution and total electric charge. For particles which can be modelled using the spherical particle model of Chapter 6, the steady state gravitational mass density solutions are similar and the total gravitational masses can be identical. This ensures that the particle and anti-particle have the same rest inertial masses.

Appendix H
qq̄ production in e⁺e⁻ collisions

H.1 $q\bar{q}$ production in e^+e^- collisions

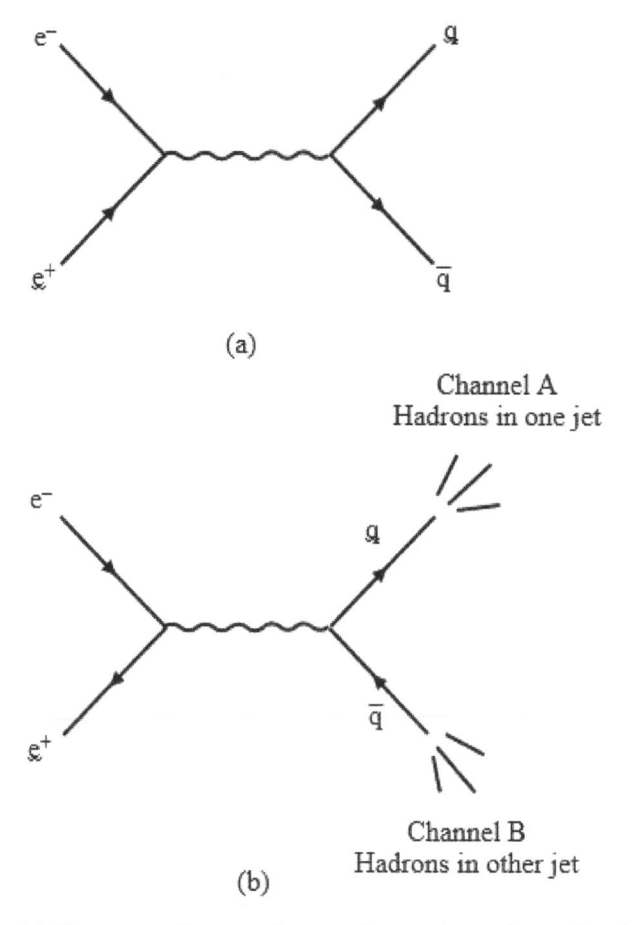

Figure H.1 (a) Feynman diagram for quark – anti-quark production from an e^+e^- collision. (b) conventional diagram for hadron production in the two output jets

$q\bar{q}$ production in e+e- collisions

Diagrams like Figure H.1(b) are shown in text books (e.g. Krane p276 (1988), Halzen and Martin p230 (1984)) whereas Cottingham and Greenwood p16 (2007) say that the precise details of the processes involved in the production of hadron jets are not yet fully understood and that there is the effect of the original quark and anti-quark pair combining with quark-anti-quark pairs generated from the vacuum to be taken into account.

Let's consider what may be involved. First of all, Figure H.1(b) cannot be correct. The algebraic sum of the charges in channel A is not $\pm ne$ where $-e$ is the electronic charge and n is a positive integer including zero. Neither is this the case in channel B. So the quarks can separate only if extra quarks are added.

We suggest that the Feynman diagram Figure H.2 applies.

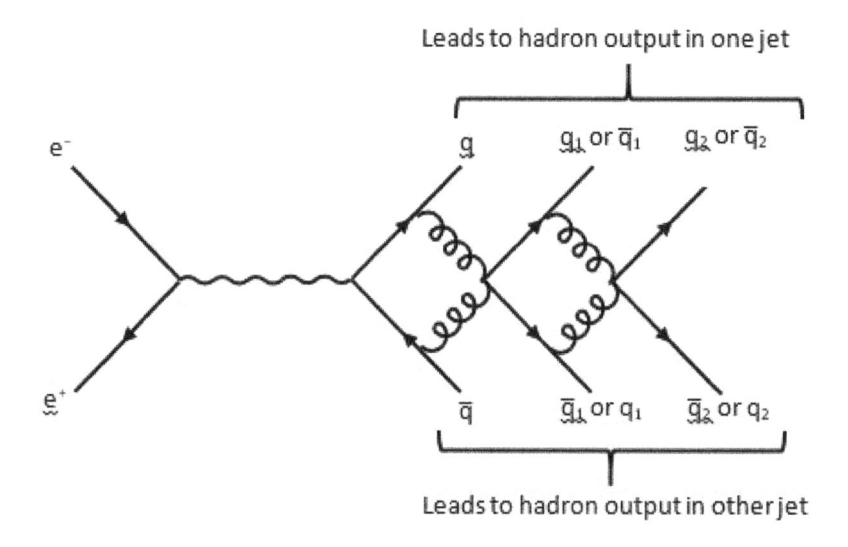

Figure H.2 Modified Feynman diagram for the proposed mechanism underlying hadron production from e^+e^- collisions. The initial $q\bar{q}$ pair cannot separate, and overlap of one by the other results in gluons and the production of a further pair $q_1\bar{q}_1$. The mechanism may be repeated with further input to the two jets, and repeated within each jet, until the final hadron composition is achieved. This ensures that there is integer charge in units of e in each jet and in each hadron

The initial $q\bar{q}$ pair cannot separate because they cannot have

Appendix H references

independent motions. They contain all the system energy. The only thing that can happen is for they to have internal motion which results in overlap. If this is incoherent, then a gluon or gluons are produced. The gluons can propagate and generate another quark – anti-quark pair, $q_1\bar{q_1}$. The process could stop here with the production of mesons or it can continue and a further $q_2\bar{q_2}$ pair formed. Again the process could stop here with the production of baryons, or it could continue. The process can also apply to further pair production within each jet. Thus there are many ways in which the $\pm ne$ condition is satisfied.

We conclude that extra quarks are required to be created to ensure that each original quark is incorporated in a particle with a charge of $\pm ne$ in each output channel. These output particles also satisfy the subset volume requirement, which the original quarks on their own were unable to do. Gravitational mass and charge configuration matching is the concept that in a decay or a particle interaction the products sum to gravitational mass and charge distributions which matches the input distributions. Qualitatively this is consistent with gravitational mass matching and hence energy conservation. This has already been invoked in the Book (Chapter 12, p265) and in Chapter 14, Section 14.5 of this book to account for the short-lived cylindrical neutral kaon form decaying to two pions, and the longer lived spherical form decaying to three pions. Here we match ring radius, electric charge and gravitational mass, and conceptually we can track individual quarks from creation to incorporation in output particles. So we conclude that in all these processes the quarks are conserved. Hence we conclude that since the output particles can only contain quarks which are in or derived from the 940 MeV set, then the first quark pair formed by the e^+e^- collision (the initial quarks) must be in the 940 MeV set. This conclusion is important because it allows the identification of the quarks which contribute to the total scattering cross section discussed in the main text, Sections 15.4 and 15.10.

Appendix H references

Cottingham W N and Greenwood D A 2007 *An introduction to the standard model* Second Edition Cambridge
Halzen F and Martin A D 1984 *Quarks and Leptons* Wiley
Krane S K 1988 *Introductory Nuclear Physics* Wiley

Appendix I

Models for the W and Z particles

I.1 Introduction

The purpose of this appendix is to provide an analysis of a proposed model for W and Z particles. In order to achieve a particle which makes use of the photon type waveforms of Chapter 3 and which is localised as a particle which can be stationary or have a velocity less than the speed of light, we propose that the radiation propagates round a circular path within the particle. The analysis is based on a model which treats the particles as members of the subset. However at the angular frequencies involved with these particles, we have to ensure that the steady state charge density is greater than the oscillatory charge density, see Chapter 6, Section 6.8.

I.2 W and Z particle models analysis

We consider Type A waveforms (see Chapter 3, Section 3.2). If we bend the Hermite function/carrier around a cylinder as described above, because it is a Type A waveform, the central section of the oscillatory charge density waveform is zero. However this is not sufficient to ensure that the net charge in the central section is zero. However if A_ρ is positive for $v = 1$, and A_ρ is negative for $v = -1$, the central steady state charges cancel and the Hermite function end sections give rise to two rotating charge spikes (either both positive or both negative) and this ensures that the steady state charge density, ρ_B, locally is greater than $1/A_\rho$. We start by examining a solution of the Field equations where there is propagation of waveforms along ϕ in cylindrical co-ordinates. We have two counter rotating rings. The Field Equations are,

$$\frac{\partial^2 m_A}{\partial r_c^2} + \frac{1}{r_c}\frac{\partial m_A}{\partial r_c} + \frac{\partial^2 m_A}{\partial z^2} + \frac{1}{r_c^2}\frac{\partial^2 m_A}{\partial \phi^2} - \frac{1}{c^2}\frac{\partial^2 m_A}{\partial t^2} = -\frac{\rho_A}{\varepsilon_0}$$

I.2 W and Z particle models analysis

$$\frac{\partial^2 \rho_A}{\partial r_c^2} + \frac{1}{r_c}\frac{\partial \rho_A}{\partial r_c} + \frac{\partial^2 \rho_A}{\partial z^2} + \frac{1}{r_c^2}\frac{\partial^2 \rho_A}{\partial \phi^2} - \frac{1}{c^2}\frac{\partial^2 \rho_A}{\partial t^2} = \frac{m_A}{\varepsilon_0}$$

We set up trial solutions along the lines of those of Chapter 3. We take the solutions to be independent of z, and in the form of a product of a function of r_c and functions of $r_0\phi$, t where r_0 is a constant.

$$m_A = \sum_n R_m(r_c) a_n h_n(r_0\phi - ct)\exp(i\omega t - ik_{mn}'r_0\phi)$$

$$\rho_A = \sum_n R_\rho(r_c) b_n h_n(r_0\phi - ct)\exp(i\omega t - ik_{\rho n}'r_0\phi)$$

where

$$\omega'^2/c^2 - k_{mn}'^2 = N_{mn}\alpha$$

$$\omega'^2/c^2 - k_{\rho n}'^2 = N_{\rho n}\alpha$$

(I.1)

On substitution and on separation of the variables we obtain

$$\frac{\partial^2 R_m(r_c)}{\partial r_c^2} + \frac{1}{r_c}\frac{\partial R_m(r_c)}{\partial r_c} + N_{mn}\alpha R_m(r_c) = 0$$

(I.2)

$$\frac{\partial^2 R_\rho(r_c)}{\partial r_c^2} + \frac{1}{r_c}\frac{\partial R_\rho(r_c)}{\partial r_c} + N_{\rho n}\alpha R_\rho(r_c) = 0$$

(I.3)

In equation (I.3) we can neglect the $N_{\rho n}$ term, see below, and the solution is proportional to $\ln(r_c/r_0)$ which can be approximated as constant in the vicinity of radius r_0. In (I.2), retaining the N_{mn} term and dropping the $1/r_c$ term, leads to a sinusoidal solution which can be arranged to a good approximation to be constant in the vicinity of radius r_0. This requires that

$$\frac{\omega_0^2}{c^2} \gg N_{mn}\alpha \gg \frac{1}{r_c^2}$$

which we shall see below is the case. Hence we need no longer concern ourselves with the radial functions. The thickness of each ring is far less than the radius (we quantify this below) and so we can put

Models for the W and Z particles

$r_c = r_0$ when separating out equations involving ϕ. Putting

$$r_0\phi' = r_0\phi - ct$$

and using

$$\frac{1}{r_0}\frac{\partial h_n}{\partial \phi} = \frac{1}{r_0}\frac{\partial h_n}{\partial \phi'}$$

$$-\frac{1}{r_0}\frac{\partial h_n}{\partial t} = \frac{1}{r_0}\frac{\partial h_n}{\partial \phi'}$$

we obtain

$$i\sum_n 2R_{mn}\left(\frac{\omega'}{c} - k'_{mn}\right)a_n \frac{dh_n}{r_0 d\phi'}\exp(i\omega't - ik'_{mn}r_0\phi')$$

$$= -\sum_n \alpha R_{\rho n} b_n h_n \exp(i\omega't - ik'_{\rho n}r_0\phi')$$

$$i\sum_n 2R_{\rho n}\left(\frac{\omega'}{c} - k'_{\rho n}\right)b_n \frac{dh_n}{r_0 d\phi'}\exp(i\omega't - ik'_{\rho n}r_0\phi')$$

$$= \sum_n \alpha R_{mn} a_n h_n \exp(i\omega't - ik'_{mn}r_0\phi')$$

If we put $z' \equiv r_0\phi'$ these last two equations are analogous to equations (3A.7) and (3A.8) Section 3A.2 in the Book, and allow us to proceed to the results obtained in Section 3A.3,

$$c_{n0} = \pm a_{n0}(N_{m0}/N_{\rho 0})^{1/2}$$

$$v_0 = \pm(N_{m0}N_{\rho 0})^{1/2}$$

and we put

$$N_{m0}N_{\rho 0} = 1$$

where we have chosen $v_0 = 1$, see (3A.15) in the Book. We require, with the particle in the subset,

I.2 W and Z particle models analysis

$$\left|\frac{m_A}{\rho_A}\right| = \left(\frac{A_\rho A}{A_m q}\right)^{1/2} \omega_0^{1/2}$$

(I.4)

and we estimate this ratio, using (3A.15) in the Book, by putting

$$\left|\frac{m_A}{\rho_A}\right| = \left(\frac{N_{\rho 0}}{N_{m0}}\right)^{1/2}$$

and we obtain

$$N_{\rho 0} = \left(\frac{A_\rho A}{A_m q}\right)^{1/2} \omega_0^{1/2}$$

(I.5)

$$N_{m0} = \left(\frac{A_m q}{A_\rho A}\right)^{1/2} \omega_0^{-1/2}$$

(I.6)

It turns out when we quantify these expressions that $N_{m0} \gg N_{\rho 0}$. We require

$$2\pi r k'_{\rho 0} = 2\pi s$$

(I.7)

$$2\pi r k'_{m0} = 2\pi (s - 1)$$

(I.8)

where s is an integer. This ensures that the carrier is continuous for both the gravitational and charge density waveforms. However the phase condition (3A.11) in the Book will not be satisfied over the whole circumference but it can be satisfied over portions of it. s is estimated below to be around 100,000. So we can have a number of large n waveforms around the circumference of the cylinder. The charges in the two rings add for $W^\pm$ particles, and the charges cancel for Z particles. We also ensure that the net angular momentum is $\hbar$ by having a reversal of the spin in going from one ring to the other. This requires that the inner ring, with radius r_2, has

$$2\pi r_2 k'_{\rho 0} = 2\pi (s - 1)$$

$$2\pi r_2 k'_{m0} = 2\pi (s - 2)$$

Models for the W and Z particles

and using

$$k'_{\rho 0} = \frac{\omega'}{c} = \frac{\omega_0}{2c}$$

so that $r - r_2 = 2c/\omega_0$ and the difference in angular momentum between the two rings is $(\hbar\omega_0/2c)(r - r_2) = \hbar$. Using

$$\frac{\omega'^2}{c^2} - k'^2_{m0} = N_{m0}\alpha$$

and making an approximation

$$\frac{\omega'}{c} - k'_{m0} = \frac{N_{m0}\alpha c}{2\omega'}$$

From (I.7) and (I.8)

$$r(k'_{\rho 0} - k'_{m0}) = 1$$

that is

$$r\left(\frac{\omega'}{c} - k'_{m0}\right) = 1$$

and so

$$\frac{rN_{m0}\alpha c}{2\omega'} = 1$$

and

$$\omega_0 = rN_{m0}\alpha c$$

and

$$\omega_0^{3/2} = r\alpha c \left(\frac{A_m q}{A_\rho A}\right)^{1/2}$$

If the $W^{\pm}$ can convert to an electron or positron plus the appropriate neutrino, then on appeal to the principle of configuration matching, we require the radius to be that of the electron ring i.e. 385 fm and this leads to a particle inertial mass $\times c^2$ of 119 GeV.

I.2 W and Z particle models analysis

$$r = \frac{1}{ac}\left(\frac{A_\rho A}{A_m q}\right)^{1/2} \omega_0^{3/2}$$

The volume is that of two counter rotating rings, each with a radial thickness of c/ω_0, and with a cylinder length l, so that

$$V = \frac{4\pi r l c}{\omega_0} = \frac{V_P}{K_P}\left(\frac{\omega_0}{\omega_{0P}}\right)^{1/2} = 2.48 \times 10^{-43} \, m^3$$

and so $l = 31$ fm. From (I.5) and (I.6) $N_{\rho 0} = 7.2 \times 10^{-20}$ and $N_{m0} = 1.4 \times 10^{19}$. Note that $\alpha N_{m0} = 1.6 \times 10^{30}$ is much less than $\omega_0^2/c^2 = 3.6 \times 10^{35}$ and so from equation (I.1) k'_{m0} is real. αN_{m0} is also much greater than $1/r_0^2 \cong 7 \times 10^{24}$ as required in solving (I.2). Note also that s is around 1.2×10^5.

Appendix J
Types of particle

J.1 Introduction

The purpose of this appendix is to pull together the various ways in which particles are categorised, in the standard model, in the Book and in the present book.

J.2 Particles in the standard model

Accounts of the stable of particles in the standard model appear in many publications along the following lines. The particles of the standard model consist of the leptons (electron, electron neutrino, muon, muon neutrino, tau, tau neutrino and their anti-particles) and quarks which exist with six different flavours (up, down, charm, strange, top, bottom and for each there is an anti-particle). Each quark can exist in any of three colour states. Quarks account for the content of baryons and mesons. Baryons contain three quarks and mesons contain a quark and anti-quark, not necessarily of the same flavour. Hadrons is the generic term for quark systems and includes all baryons and mesons. The quanta of the electromagnetic interaction are the photons, the quanta of the strong interaction are the gluons, and the quanta of the weak interactions are the $W^{\pm}$ and Z bosons. The standard model does not include gravity or a particle associated with the gravitational interaction.

J.3 Particles in the new theory

Quarks figure in the new theory. Composite forms of mesons are composed of two quarks and baryons of three quarks. Leptons (electron, electron neutrino, muon, muon neutrino and their anti-particles) are not composed of quarks. Photons are dealt with in Chapter 3 and they involve Type A waveforms. Type B waveforms,

J.4 Types of structure

Chapter 3, are the waveforms associated with gravitational-gravnetic interactions. Gluons are introduced in Chapter 15 and they arise through the mixing occurring with overlap of quarks or in the form of the neutral parton. The $W^\pm$ and the Z bosons are modelled, see Chapter 16 and Appendix I. Photons and gluons do not feature in the categories below, but the $W^\pm$ and the Z bosons do. Two separate ways of looking at the categorisation of particles follow. The first looks at the origin of the structure of particles, the other is to do with how particles interact with photons and how the expression for $\hbar$ arises in each particle case.

J.4 Types of structure

We distinguish between single omega solutions and composite structures.

Single omega solution particles These result from the analysis of Chapter 2 leading to the categories of section 2.6, Groups A, B, C and D, according to the value of the spin parameter. Groups A and C apply to mesons, Group B to leptons and baryons and Group D to baryons. In spherical or cylindrical shell solution particles of Chapter 6, all shells share a common oscillatory waveform and all parts of the particle are at the same angular frequency. There is a common spin function throughout the particle. Thus for a Group B baryon the particle has a spin of ½, and the individual shells do not have their own independent angular momentum. A quark is a single omega solution particle i.e. a single shell (see Section 14.3). Leptons have single omega solutions only.

Composite particles. These are baryons and mesons composed of quarks.

(1) Excited states of nucleons, introduced in Chapter12 in the Book, and dealt with in Chapter 14 of this book and Appendix G. The neutral parton shares around 50% of the momentum in collisions.

(2) Intermediate states (required for the formation of single omega solution states) e.g. charged kaons, Chapter 14 and Appendix G.

(3) Long lived states e.g. neutral kaons, Chapter14 and Appendix G.

(4) quark-anti-quark ($q\bar{q}$) meson with overlap potential energy.

Types of particle

These are below the volume lines on the x-energy diagrams, see Chapter 15, Section 15.5 and are introduced via the analysis of the phi meson in Section 15.6.

(5) Particle with potential energy due to overlap of quarks, with orbital angular momentum (not investigated).

J.5 Types of interaction

The categorisation resulting from this section and the following one are summarised in Figure J.1. Interactions involve two aspects The first is that they all share the same value of $\hbar$. These are the particles of the interaction set. Note that in Chapter 4 of the Book Section 4.4 it states that there may be other sets of particles interacting with a different values of $\hbar$. However in Chapter 13 in the Book it is suggested that there are no fundamental constants, but only values which arise from the adoption of a particular scheme of units. This means that the value of $\hbar$ is determined and can only have one value. Therefore there are no other sets of interacting particles. The second aspect is that in order for charged particles to interact with photons they must have the same magnitude of charge e, see Chapter 3 Section 3.5. 'All particles' in Figure J.1 include the quarks with fractional charges, confined to the interior of composite particles, but do not include Type A photon waveforms and Type B gravitational-gravnetic waveforms or gluons.

The subset is introduced in the Book, Chapter 4, Section 4.4. The sub-set is divided into the charged sub-set and the neutral subset, see Appendix 9A in the Book. The definition of the subset is discussed below but it is sufficient to say that the particles in the subset obey the subset equations of Appendix D, Section D.5. The variation of subset parameters are shown in Figures D.3 to D.9. However there are particles to which the analysis for $\hbar$ in Appendix 4A in the Book does not apply and they are not included in the subset. They include the electron, positron, electron neutrino and $q\bar{q}$ mesons.

```
                          ┌─────────────┐
                          │All particles│
                          └──────┬──────┘──────┐
                                 │      ┌──────┴──────┐
                                 │      │   Quarks:   │
                                 │      │   down up   │
                                 │      │strange charm│
                                 │      │   bottom    │
                                 │      └─────────────┘
                          ┌──────┴──────┐
                          │ Interaction │
                          │     set     │
                          └──────┬──────┘
         ┌───────────────────────┼──────────────────────┐
┌────────┴─────────┐      ┌──────┴──────┐       ┌───────┴──────┐
│Subset of charged │      │ Electrons,  │       │ Large signal │
│and neutral single│      │ positrons:  │       │  ℏ particles │
│omega solution and│      │  parameter  │       └───┬──────┬───┘
│composite particles;      │  equations  │           │      │
│parameter equations       │  Chapter 7  │      ┌────┴───┐ ┌┴──────┐
│   Appendix D     │      │   Section   │      │Electron│ │ q q̄   │
│    D11 – D14     │      │  7.3, Book  │      │neutrino│ │mesons │
└────┬─────────┬───┘      └─────────────┘      └────────┘ └───────┘
     │         │
┌────┴────┐  ┌─┴───────┐
│ Charged │  │ Neutral │
│particles│  │particles│
└─┬─────┬─┘  └─┬─────┬─┘
  │     │     │     │
┌─┴──┐ ┌┴──┐ ┌┴─┐ ┌─┴──────┐
│Muons│ │W±│ │Z │ │Neutral │
└─────┘ └┬─┘ └──┘ │hadrons │
         │        │excluding│
    ┌────┴────┐   │qq̄ mesons│
    │ Charged │   └─────────┘
    │ hadrons │
    └─────────┘
```

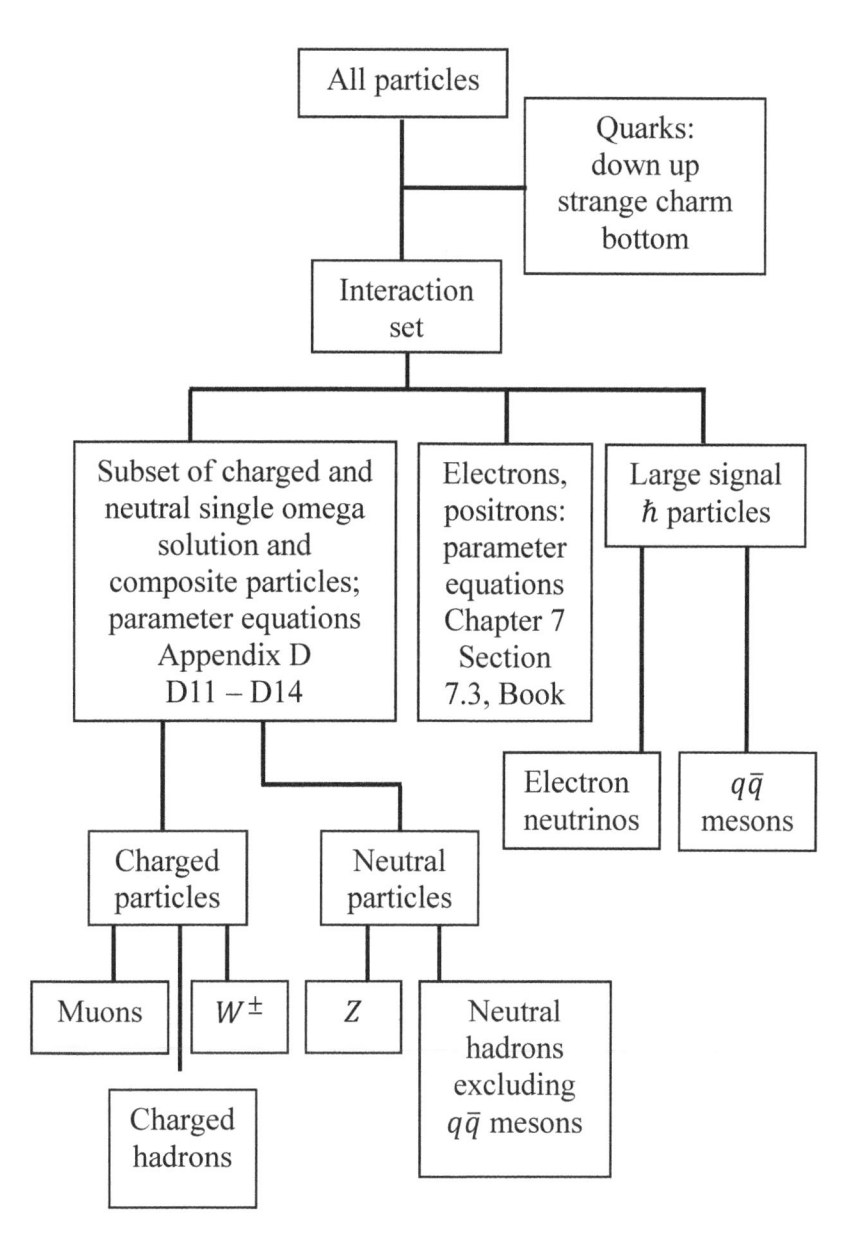

Figure J.1 Categorisation of particles within the new theory

Types of particle

J.6 The definition of the subset

In Chapter 4, Section 4.4 in the Book the subset is defined by particles having the following properties:

(1) The particle is either confined to a thin shell (or part thereof) in which the amplitudes of ρ_A and m_A are approximately constant, or m_A/ρ_A is independent of the spherical radius.
(2) There is no dependence of ρ_A on θ
(3) Defining ρ_{AM} and m_{AM} by

$$A_\rho \rho_{AM}^* \rho_{AM} \mathcal{V} = A_\rho \int_\mathcal{V} \rho_A^* \rho_A d\mathcal{V}$$

$$A_m m_{AM}^* m_{AM} \mathcal{V} = A_m \int_\mathcal{V} m_A^* m_A d\mathcal{V}$$

the particle's gravitational mass satisfies

$$|M_0| \gg |2A_\rho \rho_{AM}^* m_{AM} \mathcal{V}|$$

Given satisfaction of these conditions the analysis in Appendix 4A of the Book leading to the expression for $\hbar$ follows. It also follows that

$$\kappa \rho_{AM}^* m_{AM} = K$$

We can then see that the following are not in the subset, electrons and positrons, neutrinos (see Chapter 7 and Appendix 7C in the Book) because this condition is not satisfied, and the volume is less than that if they were in the subset. When two Dirac particles overlap (see Appendix 10A, Section 10A.9 in the Book) $\rho_{AM}^* m_{AM}$ will be large and so again the condition cannot be satisfied. So $q\bar{q}$ mesons with overlap have smaller volumes than the subset requirement.

It might be thought that because the analysis of Appendix 4A in the Book does not apply to a particular particle, the equation of energy (4.12) Section 4.5 in the Book does not apply. This is not the case. The application of Appendix 4A to subset particles allows the identification of $\hbar' = \hbar$ where $\hbar'$ is introduced in Chapter 3 in the Book. Equation (4.12) in the Book, reproduced here,

$$\frac{M_0 c^2}{(1 - w^2/c^2)^{1/2}} - M_0 c^2 + q V = 0$$

J.6 The definition of the subset

applies to all charged particles and follows from the interaction with photons, and the extension is made to include a $M_0 U$ potential energy term with the caveat that gravitation is examined further in Chapter 5 in the Book.

Appendix K
The formalism sequence

As stated in the Introduction, because the formal development of the new theory is spread over two books, an overview of the formalism sequence is required. The formalism sequence for the Book 'A Theory of Fields' and this book 'An Introduction to a Theory of Fields' is shown in Table K.1. Chapter numbers in the Book are shown in bold and chapter numbers in this book are not shown in bold. The column on the right in the table shows the various compartments of modern physics listed in the Introduction (for example, classical mechanics, special relativity mechanics, electromagnetism and so on). The new theory leads to many but not all aspects of the standard model, and this column lists the aspects included in the new theory. The main sequence in the development of the formalism is shown by the arrows running through Chapter **1** to **14** and then onto Chapters 11 to 19.

Table K.1 is largely about explaining the role of this book's appendices in extending, revising or in some cases correcting the account in the Book. Appendices F to I support Chapters 12, 14, 15 and 16. Appendix A supports Chapters **2** and **6**. Appendix D extends the content of Chapter **9**. Chapters 1 to 10 present the simplified account of Chapters **1** to **10**, but with three modifications. Chapter 3 uses a simplification of the photon waveform propagation in the background (dispersion can be neglected) explained in Appendix B. Appendix C leads to a revision of the postulates, shown in the table by feedback to Chapter **1**. As pointed out in Appendix C this does not lead to any change in the original development, because the development is essentially concerned with local situations. Chapter 18 recasts Chapter **13** in the light of the revised postulates in Appendix C. The various items in Chapter **5** (including Appendix **5B**), Chapter 5, Chapter 18 and Appendix C need to be brought together in a formal development of general relativity within the new theory. Chapter 10 supported by Appendix E presents a corrected version of Chapter **10**. There are corrections in Chapter 17 to the commandeered level of charge density in particles by gravitational potentials from distant objects presented in Chapter **12**, referred to as 'revised charge densities' in the table.

The formalism sequence

Table K.1 The formal development of the new theory

This book 'An Introduction to a Theory of Fields'			The Book 'A Theory of Fields'	Physics compartment or result
Chaps	App'ces		Chapters	
1			1	
2	A		2	General particle model
3			3	Photons
4			4	Classical mechanics, Gravity, Sp rel mechanics, Electromagnetism, Quantum mechanics
5	C		5	General relativity
6	A		6	Baryons and mesons
7			7	Leptons
8			8	Prediction of electronic charge and gravitational constant
9	D		9	Proton and neutron models
10	E		10	Nucleon scattering
			11	Nuclei models
	B		12	The background, Dirac quantum mechanics
			13	Possibility of no postulates or fundamental constants
			14	Way ahead
11				Anti-particles
12	F			Quantum mech postulates Probability
13				QFT and QED
14	G			Composite particles
15	H			Quarks, gluons
16	I			$W^\pm, Z$
17				Revised charge densities
18				Revised 'no postulates or fundamental constants'
19				Way ahead

List of symbols

Each symbol is given a title or short description – these are not definitions. The section in which the symbol is first introduced is shown in square brackets, where letters refer to the appendices. Vectors are shown in bold.

A	Magnetic potential amplitude [3.4], particle gravitational mass constant [4.3]
A_m	Steady state constant [2.3]
$A_{\theta n}$	nth component of the photon transverse oscillatory magnetic vector potential [B.3]
A_v	Amplitude of Hermite function times carrier product for gravitational mass density component [3.2]
A_ρ	Steady state constant [2.3]
A	Magnetic vector potential [3.4]
a	Electron cloud radius [13.6], Hermite function constant [B.2]
a'	$a' = c_1 + c_2 + c_3$ [9.2]
a_i	constant for ith shell [G.2]
a_n	nth component factor in travelling wave analysis [3.2]
a_{n0}	nth component factor in travelling wave analysis [I.2]
B	Constant [2.5], amplitude factor for spherical travelling wave [3.3], background factor [F.6]
B_{mA}	Constant [D.2]
B_{mA1}	Constant [7.3]
B_{mB}	Constant [D.2]
B_v	Amplitude of Hermite function times carrier product for charge density component [3.2]
$B_{\rho A}$	Constant [6.3]
$B_{\rho B}$	Constant [7.3]
b	Distance of closest approach to sun [5.5], electron torus radius of cross section [7.3], parameter in proton shell analysis [E.1]
b_n	nth component factor [3.2]
b'_ρ	Numerical constant in spherical particle model [6.3]
C	Constant [6.3], capacitance [A]
C_{mC}	Constant [D.2]

List of symbols

C_{mD}	Constant [D.2]
$C_{\rho C}$	Constant [7.3]
c	Special velocity magnitude [1.3]
c'	Radial velocity of light in gravitational field [C.2]
c_1, c_2, c_3	Charge magnitude fractions [9.2]
c_i	Charge magnitude fraction in ith shell [9.2], $|c_i|^2$ is the probability of state i [12.5]
c_{n0}	nth component factor in travelling wave analysis [I.2]
c_{Pi}	Proton charge magnitude fraction in ith shell [E.1],
D	Constant [6.3]
d	Half separation of effective point charges in pion model [6.6], capacitance plate separation [A]
d_1, d_2	Lengths of kaon shells along z axis [G.4]
d_i	Length of ith shell along z axis [14.3]
d_μ	Half length of muon cylinder along z axis [7.5]
E	Electric field magnitude [3.4], composite particle energy [14.3]
E_i	Dirac particle energy [14.3]
E_{i0}	Dirac particle rest energy [14.3]
E_P	Proton rest energy [G.4]
e	Modulus of electronic charge [8.3]
F	Force [4.4], physical function [F.3]
FSC	Fine structure constant [3.5]
f	Numerical factor in spherical particle model [6.4]
G	Gravitational field magnitude [3.4]
$\mathcal{G}$	Gravitational constant [4.4]
H	Magnetic field magnitude [3.4]
H_n	Hermite polynomial [A]
$\mathcal{H}$	Hamiltonian operator [13.2]
h	Planck's constant [4.3]
$\hbar$	$h/2\pi$ [4.3]
$\hbar'$	$h'/2\pi$ [3.5]
$h_n(z'')$	Hermite function [3.2]
i	$\sqrt{-1}$ [2.3], integer [9.2]
i_i	Dirac particle current [14.3]
J	Orbital constant [5.3]
J_0, J_1	Bessel functions [D.2]
j	Magnitude of electric current density [1.3], integer [9.2]
j_B	Magnitude of steady state electric current density [2.3]

List of symbols

j	Electric current density [1.3]
K	Gravnetic field magnitude [3.4], particle density product constant [4.3]
k	Wave vector magnitude [3.2], particle wave vector [4.2]
k'	Wave vector magnitude [3.2]
k_1, k_2	Kaon Dirac particles' wave vectors [G.4]
k_i	Dirac particle wave vector [14.3]
k_{nm}	Wave vector magnitude [B.3]
k'_{mn}, k'_{pn}	Wave vector magnitudes [3.2]
L	Dimensions of length [1.4]
l	Length of Z cylinder [I.2]
M	Particle gravitational mass [4.3], gravitational mass spike [A]
M_0	Particle gravitational mass [2.3]
M_1, M_2	Particles 1 and 2 gravitational masses [4.4]
M_1, M_2, M_3, M_4	Gravitational mass delta functions [A]
M_{i0}	Gravitational mass of ith shell [G.2]
M_{Pi0}	Gravitational mass of ith proton shell [10.3]
$\mathcal{M}_0, \mathcal{M}$	Inertial masses [4.4]
$\mathcal{M}'_0$	Inertial rest mass [F.7]
$\mathcal{M}_1, \mathcal{M}_2$	Particles 1 and 2 inertial masses [4.4]
$\mathcal{M}_{0e}$	Inertial mass of electron [8.8]
$\mathcal{M}_{0P}$	Inertial mass of proton [9.2]
$\mathcal{M}_S$	Inertial mass of sun [5.3]
$\mathcal{M}_i$	Dirac particle inertial mass [14.3]
$\mathcal{M}_{i0}$	Dirac particle rest inertial mass [14.3]
$\mathcal{M}_W$	Constant inertial mass [5.3]
m	Gravitational mass density [1.3]
m_{00}	Rest gravitational mass density [1.3]
m_A	Particle oscillatory gravitational mass density amplitude [2.3]
m'_A	Particle oscillatory gravitational mass density amplitude in background [F.6]
m_{A1}, m_{A2}	Omega waveform amplitudes [3.4], particles 1 and 2 oscillatory gravitational mass density amplitudes [10.2]
m_{AE}	Particle external oscillatory gravitational mass density amplitude [2.5]
m_{Ak}	Lepton travelling wave oscillatory gravitational mass density amplitude [7.2]

List of symbols

m_{AM} Effective mean particle oscillatory gravitational mass density amplitude [4.3]
m_{Arn} nth component of oscillatory gravitational mass density [B.3]
m_{AT} Oscillatory density associated with total steady state gravitational mass density [2.3]
m_B Particle steady state gravitational mass density [2.3]
m_{BE} Particle external steady state gravitational mass density [2.5]
m'_{BE} Reduced particle external steady state gravitational mass density [17.2]
m_{BM} Particle mean steady state gravitational mass density [4.3]
m_{BM0} Original value of particle mean steady state gravitational mass density [4.3]
m_{BT} Total steady state gravitational mass density [2.3]
m_s z direction magnetic moment quantum number [10.2]
m_{xyn} Travelling waveform nth component factor dependent on x, y [3.2]
$N_{mn}, N_{\rho n}$ Dispersion factors [B.2]
N_{EBn} Dispersion factor [B.3]
N_{EB0} Dispersion factor [B.3]
n Integer [2.3], order of Hermite function [3.2]
n_0 Central value of n for travelling wave packet [3.2]
P Algebraic sum of sinusoid amplitudes [3.2]
$\mathcal{P}$ Inertial momentum operator [13.2]
p Magnitude of gravitational momentum density [1.3]
p_{Arn} nth component of photon longitudinal oscillatory gravitational momentum [B.3]
p_{ATn} nth component of photon resultant oscillatory gravitational momentum [B.3]
$p_{A\theta n}$ nth component of photon transverse oscillatory gravitational momentum [B.3]
p_B Magnitude of steady state gravitational momentum density [2.3]
p_i Canonical variable [F.7]
$\mathbf{p}$ Gravitational momentum density [1.3]
Q Sum over sech^4 components [3.5], charge [A], Operator [F.4]
q Particle charge [2.3], observable [F.7]
q_1, q_2 Charges [4.4]
q_3 Charge [9.2]

List of symbols

q_i	Charge in ith shell [9.2], eigenvalue [F.5], canonical variable [F.7]
q_n	Eigenvalue [F.3]
q_{Pi}	Charge in ith proton shell [10.3]
R	Distance from centre of Earth [5.4], adjustment parameter at proton outer surface [9.2], ray [13.3], scattering ratio [15.10], factor in dispersion expression [B.3], operator [F.7]
R_n	Ray [13.2]
$R(r_c)$	Function of r_c [10.2]
$R_m(r), R_\rho(r)$	Functions for separation of variables [2.4]
$R_m(r_c), R_\rho(r_c)$	Functions for separation of variables in W, Z model analysis [H.2]
r	Radius in spherical co-ordinates [2.5], observable [F.7]
r_0	Parameter in spherical array analysis [3.3], parameter in W, Z model analysis [I.2]
r_{0i}	Radius at centre of ith shell [9.2]
r_{01}	Radius at centre of inner kaon shell [G.4]
r_{02}	Radius at centre of outer kaon shell [G.4]
r_1	Particle inner radius [2.5], outer spherical array radius [3.3], location of delta function [A]
r_{11}, r_{12}	Inner and outer radii of kaon inner shell [G.4]
r_2	Particle outer radius [2.5], location of delta function [A], radius of inner ring of Z model [I.2]
r_{21}, r_{22}	Inner and outer radii of kaon outer shell [G.4]
r_3, r_4	Location of delta functions [A]
r_{2P}	Proton outer radius [6.4]
r_c	Radius in cylindrical co-ordinates [6.5]
r_{c0}	Electron torus radius [7.3]
r_{ci}	Dirac particle radius [14.3]
r_{s11}, r_{s12}	Inner and outer radii of spherical kaon inner shell [G.4]
r_{s21}, r_{s22}	Inner and outer radii of spherical kaon outer shell [G.4]
r_μ	Muon cylinder radius [7.5]
S	Surface area [3.5], area contained by current loop [8.8]
S_ν	Amplitude of Hermite sinusoid [3.2]
s	Spin quantum number [2.4], integer [16.3]
T	Dimensions of time [1.4], time duration of wave packet [3.9]
t	Time [1.3]
U	Gravitational potential [2.7]

List of symbols

U_D Gravitational potential due to distant objects [17.4]
U_L Gravitational potential due to local objects [17.4]
u Orbital $1/r$ [5.3]
$\mathbf{u}$ Gravitational mass density velocity field [1.3]
V Electric potential [2.7]
$\mathcal{V}$ Particle volume [2.3] (No distinction is made between $\mathcal{V}$ and $\mathcal{V}_0$ as in the Book, Section 4.5)
$\mathcal{V}_D$ Background particle distributed volume [17.2]
$\mathcal{V}_{ni0}, \mathcal{V}_{ni}$ Neutron ith shell volume before overlap and remaining volume after overlap [10.2]
$\mathcal{V}_P$ Proton volume [6.6]
$\mathcal{V}_{Pi0}, \mathcal{V}_{Pi}$ Proton ith shell volume before overlap and remaining volume after overlap [10.2]
$\mathcal{V}_K$ Kaon volume [G.4]
$\mathbf{v}$ Charge density velocity field [1.3]
W Energy constant [4.4]
W' Energy constant [5.3]
w Velocity magnitude [4.2]
w_r Radial velocity [5.3]
X Factor [5.3], background factor [B.3], factor in kaon analysis [G.4]
x Cartesian co-ordinate [1.3], x radius corresponding to radius r [6.3] or r_c [6.5]
x' Adjusted x for proton [6.5]
x_0 x radius corresponding to radius r_0 (the shell central radius) [6.3]
x_{01}, x_{02} Kaon x_0 radii for inner and outer shells [14.5]
x_{0i} x_0 radius for ith shell [9.2]
x_{0j} x_0 radius for jth shell [9.2]
$x_1, x_2, x_3 \ldots x'_1, x'_2, x'_3 \ldots x''_1, x''_2, x''_3 \ldots$ Sequences of points [18.3]
x_{11} x radius corresponding to radius r_{11} [G.4]
x_{12} x radius corresponding to radius r_{12} [G.4]
x_{21} x radius corresponding to radius r_{21} [G.4]
x_{22} x radius corresponding to radius r_{22} [G.4]
x_{2P} Proton x radius corresponding to radius r_2 [6.5]
x_M x value, when equal to an x_{0i} at which $m_B = m_{BM}$ [D.4]
x_S Spherical particle minimum outer x radius [15.5]
x_{s01}, x_{s02} Spherical kaon x_0 radii for inner and outer shells [G.3]

List of symbols

x_π Pion sphere radius in x units [6.6]
y Cartesian co-ordinate [1.3]
$Z(z)$ Function of z [10.2]
z Cartesian co-ordinate [1.3], cylindrical co-ordinate [6.5]
z' $z - ct$ [B.2]
z'' Proportional to $z - ct$ [3.2]
z_0 Offset distance [7.2]
z_s Separation between shell centres [10.2]
$\mathbf{z_1}$ Unit vector [10.2]

α $1/\varepsilon_0$ [3.7]
β_{ni} Overlap parameter for neutron ith shell [10.2]
β_{Pi} Overlap parameter for proton ith shell [10.2]
γ Gravitational red shift parameter [5.3], variable factor [A]
Δk Photon wave vector magnitude [3.4], shift in wave vector [4.3]
ΔM Difference in gravitational mass [A], transported gravitational mass [B.3]
$\Delta r_c, \Delta r_{c0}$ Electron extent in r_c direction [7.3], quark ring thickness [15.2]
Δr_μ Muon shell thickness [7.5]
Δx Shell half thickness in x [14.3]
Δx_i Shell half thickness in x for ith shell [D.4]
Δx_{max} Maximum shell half thickness in x [6.4]
Δx_{s1} Shell half thickness in x for spherical kaon inner shell [G.4]
Δx_{s2} Shell half thickness in x for spherical kaon outer shell [G.4]
Δx_μ Muon cylinder thickness in x units [6.6]
Δx_π Pion spherical shell thickness in x units [6.6]
$\Delta z, \Delta z_0$ Electron extent in z direction [7.3]
$\Delta \theta$ Proton and neutron angle [6.5]
$\Delta \theta_3, \Delta \theta_{3max}$ Proton angles [9.2]
$\Delta \phi$ Deflection angle of radiation in a gravitational field [5.4]
$\Delta \omega$ Photon angular frequency [3.4], angular frequency difference [4.4]
ε_0 Permittivity of free space [1.3]
ε_i Sign of charge in the ith shell [G.2]
η Proton volume adjustment parameter [D.3]
$\Theta_m(\theta), \Theta_p(\theta)$ Functions for separation of variables [2.4]
θ Colatitude in spherical co-ordinates [2.4]

List of symbols

κ	Parameter for taking account of positively and negatively charged particle shells [15.5]
κ_P	Parameter for taking account of positively and negatively charged proton shells [15.5]
κ'_P	Parameter for taking account of positively and negatively charged proton shells [D.2]
μ	Infrared cutoff [13.6]
μ_e	Electron magnetic moment [8.8]
μ_P	Proton magnetic moment [9.2]
μ_0	Magnetic permeability of free space [8.2]
ν	$n - n_0$ [3.2], scattered wave [12.3]
ν_0	Constant in travelling wave analysis [3.2]
ν_1, ν_2, ν_3	Particle circulation parameters [9.2]
ρ	Electric charge density [1.3]
ρ_{00}	Rest charge density [1.3]
ρ_A	Particle oscillatory charge density amplitude [2.3]
ρ'_A	Particle oscillatory charge density amplitude in background [F.6]
ρ_{A1}, ρ_{A2}	Particles 1 and 2 oscillatory charge density amplitudes [10.2]
ρ_{A2E}	Particle external oscillatory charge density amplitude [2.5]
ρ_{AD}	Particle oscillatory charge density due to distant objects [17.4]
ρ_{Ak}	Lepton travelling wave oscillatory gravitational mass density amplitude [7.2]
ρ_{AL}	Particle local oscillatory charge density [17.4]
ρ_{AM}	Effective mean particle oscillatory charge density amplitude [4.3]
ρ_{AM0}	Original value of effective mean particle oscillatory charge density amplitude [4.3]
ρ_B	Particle steady state charge density [2.3]
ρ_{B1}	Internal component of particle steady state charge density obeying Laplace's equation [2.5]
ρ_{B2}	Internal component of steady state charge density obeying Poisson's equation [2.5], charge density in centre of outer shell [G.4]
ρ_{B22}	Steady state charge density at outer surface of outer shell [G.4]
ρ_{B2E}	External continuation of ρ_{B2} [2.5]
ρ'_{B2E}	Reduced external continuation of ρ_{B2} [17.2]
ρ_{BE}	Particle external steady state charge density [2.5]

List of symbols

ρ_{Bi}	Steady state charge density at centre of ith shell [9.2]
ρ_{Bi0}	Original steady state charge density at centre of ith shell [10.3]
ρ_{Bj}	Steady state charge density at centre of jth shell [9.2]
ρ_{BM}	Particle mean steady state charge density [4.3]
ρ_{BM0}	Original particle mean steady state charge density [4.3]
ρ_{xyn}	Travelling waveform nth component factor dependent on x, y [3.2]
σ	Scattering cross section [15.10]
Φ, Φ'	Gravitational potential based on inertial mass [C.2, C.4]
$\Phi_m(\phi), \Phi_\rho(\Phi)$	Functions for separation of variables [2.4]
$\Phi_{m0}, \Phi_{\rho 0}$	Constants [2.4]
ϕ	Spherical co-ordinate [2.4], cylindrical co-ordinate [6.5]
ϕ'	Related to $r_0\phi - ct$ [I.2]
ψ	Wave function [F.5]
ψ_a, ψ_b	Eigenstate wave functions [F.4]
ψ_D	Wave function [4.2]
ψ_i, ψ_j	Eigenstate wave functions [F.5]
ψ_n	Eigenstate wave function [F.3]
ψ_R	Radial wave function [12.3]
ψ_T	Translational wave function [12.3]
ψ_W	Potential well wave function [12.3]
ψ_{Wi}	Potential well wave function for state i [F.6]
ψ_P	Probabilistic wave function [12.3]
ψ_{P1}	Probabilistic wave function [12.3]
ψ_{P2}	Probabilistic wave function [12.3]
ψ_{Pi}	Probabilistic wave function for state i [F.6]
Ψ, Ψ_n	Vectors [13.3]
ω	Travelling wave angular frequency [3.2], particle angular frequency [4.2]
ω'	External field half angular frequency [3.2], modified angular frequency [C.4]
ω_0	Particle angular frequency [2.3]
ω_0'	Particle rest angular frequency [F.7]
ω_{00}	Lepton angular frequency constant [7.2]
ω_{0e}	Electron angular frequency [7.3]
ω_{0P}	Proton angular frequency [6.5]
$\omega_{0\mu}$	Muon angular frequency [7.5]

List of symbols

$\omega_{0\pi}$ Pion angular frequency [6.6]
ω_1 Omega waveform angular frequency [3.4], quark angular frequency [15.6]
ω_2 Omega waveform angular frequency [3.4], quark angular frequency [15.6]
ω_i Dirac particle angular frequency [G.2]
ω_{i0} Dirac particle rest angular frequency [14.3]
ω_κ Kaon angular frequency [G.4]

Index

Some items are used throughout the book and they are only indexed by the earliest mentions.

Accuracy of predictions 201-202
Advance of the perihelion 55
Ampere's law xiii, 47
Anti-hydrogen 118
Anti-particles 13, 90, 131, 132:
anti-proton 70
inertial mass 271-272
kaons 263
positron 79
with negative gravitational mass 115-117
with positive gravitational mass 117-118
Asymptotic freedom 178
Atomic physics xiii, 49
Atoms 8, 40, 49, 135, 190, 199, 219

Background, the 21-22, 187-190:
background particles 190
dark energy 204
dark matter 193, 204
oscillatory mechanism for all possible outcomes 202, 256-257
propagation of omega waveforms 28-30, 216
propagation of photons 31, 213-216

spherical array 29-30
transmission of potentials 27-30
Baryons 13-14, 59, 164-166, 282-283, 285
Basic particle 43
Bending of light in a gravitational field 56
Beta decay 182-184
Bottom quark 173-174, 179, 282
Bottomonium 173

Canonical variables 134, 259
Cartesian axes 3
Charge *see* electric charge
Charge conjugation 132
Charge density *see* electric charge density
Charm quarks 169-173, 177, 282
Charmonium 173
Classical mechanics 18, 40, 45-48
Colliders 163
Colour 162, 178, 201, 282
Complete sets 254
Compartments of physics xii, 288
Composite particle 144-154, 261-262, 283
Constant K 69, 87
Cosmology 204

Index

Coulomb's law xii, 6, 47
Cross-sections 122-124, 131, 177, 275
Current density 4

Dark energy 204
Dark matter 204
Delta omega waveform 213
Descartes 195
Dipole radiation 33, 34-35
Dirac particle 140-143
Dirac relativistic quantum mechanics 129-130
Discontinuities in charge density 206-208
Dispersion relationship for particle wave packet 41
Dispersion relationships in omega and delta omega waveform analysis 212, 213
Distributed particle 43
Double slit experiment 125, 255-257
Down quark 118, 164, 177, 282

e^+e^- collisions 134, 163-164, 175-177, 274-275
e^+e^- creation 116, 133-137
Earth 50, 55, 221
Eigenvalue 252-253
Eigenvalue equation 252
Einstein xiii, 2, 50
Electric charge 13, 36:
charge in spherical and cylindrical shells 234-235
Electric charge density 4:
charge densities due to distant objects 192-193

oscillatory 9-10, 23-24, 44-45
rest value of charge density 6, 12
steady state 9-10, 30-32
Electric current density 4, 10-11
Electric dipole moment 74
Electric field 32-34, 46
Electric potential 20-21, 27-30, 43-46, 133:
relation to gravitational mass density 20
Electric potential energy 45-46, 48
Electromagnetic radiation 30-38, 216:
and particle motion 221-222
Electromagnetism xii, 32, 48, 133
Electron 76-79, 182:
angular frequency 84
cloud radius 136
energy 84
magnetic moment 89, 135
model 76-79, 190-191
structure 76-79, 88-89
torus 77
Electron-positron pairs 136-137
Electron neutrino 80-81, 116, 190-191, 282, 284
Electronic charge 85, 232
Electroweak theory 181-183, 184, 203-204
Energy:
preliminary introduction 1, 34-35
formal introduction 46
Equation of energy 21, 45, 52
Equation of motion 46
Expectation value 254-255
External potentials 20-21

301

Index

Falsifiability 195, 198-189
Faraday's law of induction xiii, 47
Feynman diagrams 133-134, 175-176, 273
Field 1-2
Field equations, 3-6
Field quantisation 24
Fine Structure Constant 35-36, 85-86
Force 46
Formalism sequence of new theory 288
Formalisms xii-xiii
Frequency balance equation 53
Fundamental constants 12, 44, 83, 199, 204-205

Galaxy 192
General relativity xiii, 40, 50, 55, 56, 57-58, 204, 220, 225
Gluons 162-163, 168-169, 178-179, 275
Gravitational analogue of Ampere's law 47
Gravitational constant 47, 87-88, 98-101, 201-202, 232
Gravitational – gravnetic Type B waveform 37, 50
Gravitational mass :
steady state gravitational mass 12
steady state gravitational mass in spherical and cylindrical shells 234-235
transport of gravitational mass 36-37
Gravitational mass and charge configuration matching 132, 151, 153, 179, 275

Gravitational mass density 3:
rest value of gravitational mass density 6, 11
oscillatory 9-10, 24, 43-44
steady state 9-10, 14, 18-20, 20, 44
Gravitational momentum density 3, 10, 32
Gravitational potential 21, 43-44, 45, 48:
relation to charge density 21
Gravitational potential energy 44, 48
Gravitational red shift 55-56
Gravitational red shift parameter 52, 55
Gravnetic field 47
Group velocity 42, 221

Hadrons 59, 177-8, 282
Hamiltonian operator 129, 133, 252-253, 257-259
Heisenberg representation 130
Hermite function 24-25, 212
Hermitian operators 253-254
Hidden variables 121
Higgs field 181-182
Higgs mechanism 182
Higgs particle 181-182
Hilbert space 130

Impact parameter 122, 124
Inertial mass 45:
rest inertial mass 46-47
Inertial momentum 46
Inertial momentum operator 129
Infrared cutoff 136
Interaction representation 130

Index

Interaction set of particles 284-285
Interference:
constructive 107
destructive 107

J/ψ particle 173
Jets 163-164

Kaon, 72, 146-153, 262-263:
charged 147-149, 153, 263-268
decay 149-153
neutral 149-153, 269-270
quark content 158-162
Kinetic energy 45
Klein-Gordon equation 130

Lagrangian, relativistic 134
Lamb shift 135-137
Lambda decay 116
Laplace's equation 15, 61, 117
Leptons 13, 75, 282
Lepton standing wave solution 76
Lorentz gauge 133
Low frequency filtering 30-31

Magnetic field 32, 47, 135, 213
Magnetic moments 71-72, 89, 96-101, 135
Magnetic vector potential 31-32, 47, 133, 213-215
Matrix representation 130
Maxwell's equations 32, 133
Measurement 125, 252-253
Mesons 13-14, 72-74, 165-166, 282
Mixing 30, 106-107, 168-169
Multi-shell particle model 61-66, 206

Muon 79-80, 282, 285:
decay 116
structure 89
production 134, 175-176
Muon neutrino 116, 282
Muon neutrino velocity 57, 220-222

Near-compensation charge 188-191
Near-compensation gravitational mass 188-191
Neutral particles 70-72
Neutral parton 144-146, 283
Neutrinos 76, 80-81, 118, 282:
neutrino rest energy 89-90
Neutron:
application of the multi-shell model 70-72
charge fractions 101, 103
circulation parameters 101
cylindrical shells 103
magnetic moment 101
quark content 157-162
three shell structure 101,103
Newton's law of gravitation 47, 87-88
Newton's laws of motion 40, 46
Nuclear models 114, 202
Nucleon spin 108
Nucleon interactions 105-114, 241-249
Nucleus 114, 202

Observable 252
Occam xii
Occam's razor xxii, xiii
Omega waveform 23-30, 211-213

Index

Operator 252, 254, 259-260
Orbital constant 55
Orbital equation of energy 54
Oscillatory components, 9-10
Oscillatory mechanism for all possible outcomes 202, 256-257
Oscillatory solutions of the Field Equations:
electron 77
general solution 13-14
omega waveforms 23-30, 211-213
pion 72
proton 66
spherical particle model 61
standing wave 76
W and Z particles 277
Overlap model for phi meson 167-169
Overlap model for the nuclear force 105
Overlap parameter 106
Overlap potential energies 109-114, 241-249

Parity 132
Particle:
angular frequency 9-10
categories 13-14
charge 13
decays 116, 149-152, 203
densities versus angular frequency 60-61, 235-240
geometrical configuration 88-91
gravitational mass 12
interactions 203
interaction set 36, 284, 285
internal fields oscillatory 9-10
internal fields steady state 9-10

model 9
near-compensation charge 188-191
near-compensation gravitational mass 188-191
parameters 60-61, 232, 233-234, 235-240
stability 66, 68
subset 60-61, 235-240, 261-262, 284-287
travelling wave packet 42
wave vector 41, 48
Particle density product constant 45, 69, 87, 232
Particle gravitational mass constant 45, 69, 87, 232
Particles with overlap 284:
bottomonium 173-174
charmonium 173
phi meson 167-169
Perihelion, advance of 55
Permittivity of free space 4, 83-84
Permeability of free space 84
Phi meson 167-169
Photons 30-38:
amplitude 32
difference angular frequency 31
dipole radiation 33, 34
low pass filtering 30, 31
magnetic vector potential 31-32
mixing of ω_1 and ω_2 waveforms 30-31
propagation in background 213-216
propagation in particles 31-37, 216
radiation energy transport 34-36
summary of properties 37-38

Index

transport of gravitational mass 36-37
transverse magnetic vector potential 32
with negative gravitational mass 115-117
Pion 72-74:
composite form 153-154
decay 116
predicted energy 88
quark content 18, 161
single omega solution form 153
structure 88, 91
Planck's constant 36, 44, 84, 167, 284, 286
Poisson's equation 16
Poisson brackets 252, 259-260
Popper 195
Positron 79, 136-137
Postulates xii, xiii
Postulates of the new theory 1, 2, 37, 195-196, 225
Postulates of quantum mechanics 125-128, 250-260
Potential well 257-258
Probabilistic wave function 122
Probability in quantum mechanics 121-125
Proton 66-70:
angular frequency 84
charge fractions 94-100
circulation parameters 97
cylindrical shells 94, 102, 230, 232-233
energy 84
excited states 144-146
magnetic moment 96-101
multishell model 66-68
multishell structure 67-68, 88, 91
outer x radius 65, 68, 85
partial spherical shells 226-231
quark content 158-162
stability 66, 68
Set1, Set2, Set3 97
three shell structure 94
volume 86
x-energy diagram 139-140
Pure logic sequence 197-199

Quantisation 24, 35
Quantum chromodynamics 156, 178, 203
Quantum electrodynamics 129, 133-137, 202
Quantum field theory 129-137, 202
Quantum mechanics 48-49, 120-128, 202
Quantum mechanics postulates 125-128, 250-260
Quarks 156, 157-162, 177-179, 282-283:
asymptotic freedom 178
bottom 173-174, 177, 282
charm 169-173, 282
down 159-161, 170-172, 282
fractional charges 94-95, 97, 101, 158, 160
pair production 163-164, 273-275
strange 159-162, 167-169, 170, 282
top 203, 282
up 159-161, 170-172, 282

Radiation energy 34-37
Ray (in QFT) 131

Index

Relative values of fields, charges, gravitational masses 14-18
Relativistic constraints 6, 10-11
Relativistic equation of energy 286-287
Rest energy, 46
Rest inertial mass 46-47
Rutherford scattering 122-124

Scattering ratio R 175-177
Schrödinger representations 130
Schrodinger's equation 48, 123, 257
Shell:
Dirac particle 140-143
inertial mass and energy 143
kaon 146-153, 263-270
maximum thickness 65-66
multi-shell particle model 61-65
neutron 69-72, 101,103
proton 66-68, 94-102
quarks 157-162
SI units xvii, 4, 204-205
Single omega solutions 14
Single omega solution particles 14, 139-140, 283
Solid state physics 49
Special relativity 2, 40
Special velocity magnitude 2
Spherical particle shell model 61-65
Spin angular momentum 48
Spin quantum number 13, 14
Spontaneous emission 124, 255
Standard clock 222
Standard clock in free fall 222-223
Standard model 156, 177-79, 181-182, 203-204, 282

Standing wave solution 77-76
Steady state components 10
Steady state constants 11-12
Strange quark 159-162, 167-169, 170, 282
Strong equivalence principle 225
Subset 60, 235-240, 261-262, 284-286
Subset equations 60-61, 235-236
Sun 51, 56, 192-193, 221
Symmetries 131-134, 15.11, 181-182, 203
Symmetry operators 131-132

Time dependent perturbation theory 124
Time reversal 132
Top quark 203, 282
Transition probabilities 131
Transport of gravitational mass 36-37, 45
Type A waveform 25-26, 213
Type B waveform 25

Universe 192-193, 204
Up quark 159-161, 282

Vacuum polarisation 135, 136-137
Vector space 130
Velocity field 1-2
Velocity of light 31, 83, 84
Venus 55
Virtual photon propagator 135
Volume lines on x-energy diagram 164-166, 167

$W^{\pm}$ particle model 184-185, 276-

Index

281, 282-283
Wave function 41, 43, 48-49, 121-128, 129-133, 251-260
Wave packet 42
Weak interactions 181-182

x-energy diagram 139-140 *et seq*

Z particle model 184-185, 276-281, 282-283